Jing Sun

Control Engineering

Also of Interest

Fractional-Order Control Systems
D. Xue, 2017
ISBN 978-3-11-049999-5, e-ISBN (PDF) 978-3-11-049797-7, e-ISBN (EPUB) 978-3-11-049719-9

Signals and Systems, Volumes 1+2
W. Zhang, 2017
ISBN 978-3-11-054409-1

The Fundamentals of Electrical Engineering
F. Hüning, 2014
ISBN 978-3-11-034991-7, e-ISBN (PDF) 978-3-11-034990-0, e-ISBN (EPUB) 978-3-11-030840-2

Basic Process Engineering Control
P. Agachi, M. Cristea, 2014
ISBN 978-3-11-028981-7, e-ISBN (PDF) 978-3-11-028982-4, e-ISBN (EPUB) 978-3-11-037701-9

Advanced Process Engineering Control
P. Agachi, M. Cristea, 2016
ISBN 978-3-11-030662-0, e-ISBN (PDF) 978-3-11-030663-7, e-ISBN (EPUB) 978-3-11-038816-9

Jing Sun

Control Engineering

Fundamentals

DE GRUYTER

Author
Prof. Jing Sun
Dalian University of Technology
School of Mechanical Engineering
No. 2 Linggong Rd
116024 Dalian
China

ISBN 978-3-11-057326-8
e-ISBN (PDF) 978-3-11-057327-5
e-ISBN (EPUB) 978-3-11-057336-7

Library of Congress Cataloging-in-Publication Data

Names: Sun, Jing (Computer scientist) author.
Title: Control engineering : fundamentals / Jing Sun.
Description: First edition. | Berlin ; Boston : Walter de Gruyter GmbH, [2018] | Series: De Gruyter textbook | Includes bibliographical references and index.
Identifiers: LCCN 2018007856| ISBN 9783110573268 (softcover : acid-free paper) | ISBN 9783110573275 (pdf) | ISBN 9783110573367 (epub)
Subjects: LCSH: Automatic control.
Classification: LCC TJ213 .S7988 2018 | DDC 629.8–dc23 LC record available at https://lccn.loc.gov/2018007856

Bibliographic information published by the Deutsche Nationalbibliothek
The Deutsche Nationalbibliothek lists this publication in the Deutsche Nationalbibliografie; detailed bibliographic data are available on the Internet at http://dnb.dnb.de.

Typesetting: Integra Software Services, Pondicherry
Printing and binding: CPI books GmbH, Leck
Cover image: Baranozdemir/iStock/Getty Images

www.degruyter.com

Preface

The publication of this book is motivated by two backgrounds, a major one and a subsidiary one, respectively.

First, let's talk about the major background. In order to improve the international competitive power of China's tertiary education, Ministry of Education of the People's Republic of China released a document, named "Several opinions on strengthening undergraduate teaching works and improving teaching quality of colleges and universities" in 2001. The document explicitly points out that foreign language, English in particular, should be applied in general courses and the specialized courses teaching process for undergraduate education. The document also encourages colleges and universities to publish an English version textbook or to introduce an original English textbook, especially the former. Since 2001, a number of colleges and universities have opened the specialized courses given in English. And most colleges and universities selected the original English textbook, and some colleges and universities just adopted the lecture notes written by teachers. Up till now, fewer English versions specialized textbook are published, especially about the control engineering.

Next, let's talk something about the subsidiary background. Our university, Dalian University of Technology, always encourages teachers to publish English version textbooks with the distinct specialized characteristic. There are several aspects for this encouragement. ①Although the original English textbooks are classical, they cannot match the teaching syllabus and teaching system. Hence, the original ones are not suitable for our course and profession construction. ②Some descriptions in original textbooks, such as signal, figure, equation, case, principle, and terminology, are different from those used in relevant courses givenin Chinese. The students cannot make a perfect connection with other courses so that they might be puzzled when trying to understandsome principles or terminologies. ③For most courses, the instructors are faculty of the university who have taught this course in Chinese for several semesters. They would feel comfortable and give a nice presentation with the self-written English textbook, at the same time, and they could bring their professional advantages into the classes.

With the motivation of the above backgrounds as well as the experience of auditing two courses about control theory in the United States, the author of this textbook, who is in charge of teaching the same subject in DUT, realized the urgency of releasing the English version textbook by herself. Combined with her two-year teaching experiment, the first edition of 'Fundamentals of Control Engineering' is about to get published.

The main readers of this textbook are the students with major in mechanical engineering. Based on the main content of the classical control theory, the author has referred to both several original textbooks in English and Chinese version. According to the requirement of China's national general teaching syllabus for the control

https://doi.org/10.1515/9783110573275-201

courses, our textbook provides useful and powerful aids when the readers want to solve some theoretical and practical problems in the field of classical control theory. Considering the characteristic of mechanical engineering, the author aims to state some basic concepts, knowledge, terminologies, equations, and methods for the classical control theory.

There are nine chapters in the whole book, including Introduction, Laplace Transforms Solution, Formulation and Dynamic Behavior of Translational Mechanical Systems, Formulation and Dynamic Behavior of Electrical System, Fundamentals of Control Systems, Time Response Analysis of Control Systems, Frequency Response Analysis of ControlSystems, Stability Analysis of Control Systems, and Error Analysis and Calculating of Control Systems. Our suggested in-class hours are 32 ~ 56, and the suggested in-class hours for every chapter are as the following.

Chapter	Suggested in-class hours
1 Introduction	2~4
2 Laplace Transforms Solution	2~4
3 Formulation and Dynamic Behavior of Translational Mechanical Systems	2~6
4 Formulation and Dynamic Behavior of Electrical Systems	2~2
5 Fundamentals of Control Systems	6~10
6 Time Response Analysis of Control Systems	4~6
7 Frequency Response Analysis of Control Systems	6~10
8 Stability Analysis of Control Systems	4~6
9 Error Analysis and Calculating of Control Systems	4~8

This book is written by Dr. Jing Sun of the Dalian University of Technology. Special thanks to all the graduates of Room No.305 in the School of Mechanical Engineering and to some undergraduates of the international class in the School of Mechanical Engineering. They took great effort in selecting materials and revising the entire book. They did a big favor in the publication of this book with their selfless contribution.

Acknowledgement to Professor Meng Zhang, Kansas State University, USA. The author is grateful to Professor Zhang's kindness in modifying some inappropriate expression for the entire book. Acknowledgement to Professor Zhenyuan Jia, Dalian University of Technology. He gave many guidance, suggestions and modification direction. Express my deep appreciation to Professor Yongqing Wang, Dalian University of Technology. He revised some chapters of this book.

This book is dedicated to Mr. Mingyu Xu.

Jing Sun
School of Mechanical Engineering
Dalian University of Technology
December, 2016

Contents

1 Introduction

Fundamentals of control engineering is a part of cybernetics. Cybernetics is a branch of science that focuses on the control relationships and regulation rules among different organisms, machines, and systems. Cybernetics is not only an important branch of science, but also a remarkable methodology.

Engineering cybernetics is accompanied by the combination of cybernetics and practical engineering problems. The basis of engineering cybernetics was introduced in the book, *Engineering Cybernetics*, by Xuesen Qian in 1954. In this outstanding work, the concept of engineering cybernetics was put forward for the first time and cybernetics was generalized in the field of engineering.

In fundamentals of control engineering, also known as fundamentals of control theory, we are concerned with some fundamental theories of automatically controlled technology. In fact, fundamentals of control engineering is a part of engineering cybernetics and belongs to the field of classical control theory.

So, in general, this book focuses on introducing the principles and methods in the field of classical control theory. In particular, the emphasis is on dynamics and controls of mechanical engineering systems. In addition, the importance of understanding and being able to determine the dynamic response of physics systems is the main purpose of the book.

1.1 System and System Analysis

Because the most frequently used keyword in this book is "system", we should define it at the onset. A system is a collection of interacting elements for which there are cause-and-effect relationships among the variables. This definition is certainly general, because it must encompass a broad range of systems. The most important feature of the definition is that interactions among the variables, rather than individual elements, should be taken into account in system modeling and analysis. So, a system is a combination of components or elements and is constructed to achieve an objective or multiple objectives.

Our study concentrates on dynamic systems, where the variables are time dependent. In most of our examples, not only will the excitations and responses vary with time, but also at any instant the derivatives of one or more variables will depend on the values of the system variables at that instant. The system's response will normally depend on initial conditions, such as stored energy, in addition to any external excitations.

In the process of analyzing a system, two tasks must be performed: modeling the system and solving for the model's response. The combination of these two steps is referred to as system analysis.

https://doi.org/10.1515/9783110573275-001

1.2 Modeling the System

A mathematical model, or model for short, is a description of a system in terms of equations. The physical laws are the basis for constructing a model of a system (such as the conservation of energy and Newton's laws), which the system elements and their interconnections need to obey.

The type of model sought depends on both the objective of the engineer and the tools for analysis. If a pencil-and-paper analysis with parameters expressed in literal rather than numerical form is to be performed, a relatively simple model will be needed. To achieve this simplicity, the engineer should neglect elements that do not play a dominant role in the system.

On the other hand, if a computer is available for carrying out simulations of specific cases with parameters expressed in numerical form, a comprehensive mathematical model that includes descriptions of both primary and secondary effects might be appropriate. In short, a number of mathematical models are possible for a system, and the engineer must decide which form and complexity are most consistent with the objectives and the available resources.

The most common example of a dynamic system is the automobile. To limit the complexity of any model, some of the system's features must be omitted. In fact, many of the parameters may be relatively unimportant for the objective of a particular study. Among many possible concerns are the ease of handling on the straightaway or while turning around a corner, comfort of the driver, fuel efficiency, stopping ability, crash resistance, the effects of wind gusts, potholes, and other obstacles.

Suppose that we limit our concern to focus on the driver when the vehicle is traveling on a rough road. Some of the key characteristics of the system are represented in Fig. 1.1(a) by masses, springs, and shock absorbers. The chassis has by far the largest mass, but other masses that may be significant are the front axles, rear axles, wheels, and driver. Suspension systems between the chassis and the axles are designed to minimize the vertical motion of the chassis when the tires undergo a sudden change in motion because of the road surface. The tires themselves have some elasticity, which is represented in Fig. 1.1(a) by additional springs between the wheels and the road. The driver is slightly cushioned from the chassis motion due to the characteristics of the seat, and there is also some friction between the driver and the seat-back.

Fig. 1.1 is adapted from a drawing in Chapter 42 of *The Shock and Vibration Handbook*, third edition (1988), edited by Cyril M. Harris. It is used with the permission of the publisher, McGraw-Hill, Inc. Fig. 1.1a also appears in the fourth edition (1996) of that book.

Let us assume that the vehicle is traveling at a constant speed and that the horizontal motion of the chassis does not concern us. We must certainly allow for the vertical motion caused by the uneven road surface. We may also consider the

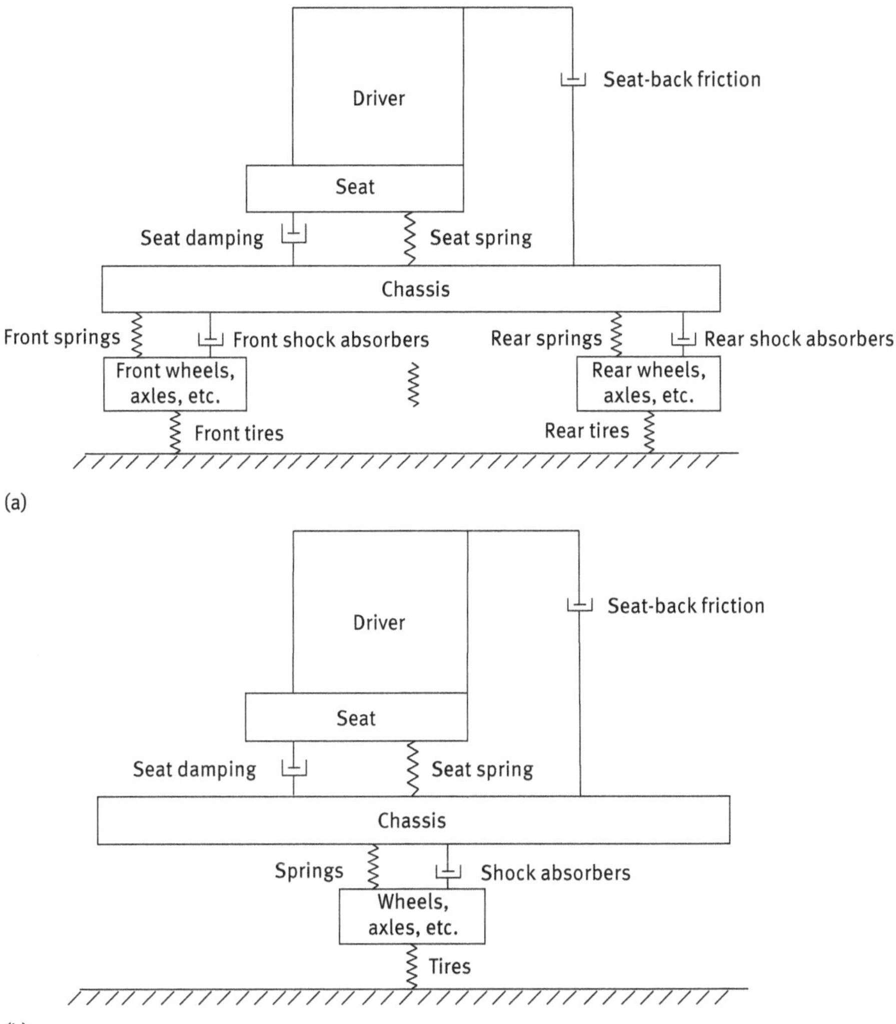

Fig. 1.1: The representation for automobile: (a) a representation of an automobile and (b) a simplified representation.

pitching effect when the front tires hit a bump or depression, causing the front of the chassis to move up or down before the rear. This would require us to consider not only the vertical motion of the chassis but also rotation about its center of mass.

The complexity of a system model is sometimes measured by the number of independent energy-storing elements. As in Fig. 1.1(a), energy can be stored in four different masses and in five different springs. If the pitching effect is ignored, the analysis might be simplified by combining the front and rear axles into a single mass, as shown in Fig. 1.1(b), which has only three masses and three springs.

In the initial phase of the analysis, other simplifying assumptions might be made. Perhaps some of the elements remaining in Fig. 1.1(b) could be omitted. Possibly a mathematical description of the individual elements that is simpler than that required for the final analysis would be used.

On the other hand, for a more thorough study of the effect of a bumpy road on the driver, it might be necessary to add other characteristics to those represented in Fig. 1.1(a).When one of the two wheels on the front axle encounters a bump or depression, the displacement and forces on it are different from those on its mate. Thus, we might want to consider each of the four wheels as a separate mass and to allow for side-to-side rotation of the chassis, in addition to the vertical and pitching motions.

When devising models at various stages in the design process, engineers usually give considerable thought to how detailed the representation of the system's characteristics should be. Many of the remarks we have made about the automobile can be extended to airplanes, boats, rockets, motorcycles, and other vehicles. In the following chapters, we shall show how to describe important characteristics by sets of equations.

1.3 Solving the Model

The process of using a mathematical model to determine certain features of the system's cause-and-effect relationships is referred to as solving the model. For example, the responses to specific excitations may be desired for a range of parameter values, as guides are selecting design values for those parameters. As described in the discussion of modeling, this phase may include the analytical solution of simple models and the computer solution of more complex ones.

The type of equation involved in the model has a strong influence on the analytical methods. For example, nonlinear differential equations can seldom be solved in closed form, and the solution of partial differential equations is far more laborious than that of ordinary differential equations. Computers can be used to generate the responses to specific numerical cases for complex models. However, using a computer to solve a complex model has its limitations. Models used for computer studies should be chosen with the approximations encountered in numerical integration in mind and should be relatively insensitive to system parameters whose values are uncertain or subject to change. Furthermore, it may be difficult to generalize results based only on computer solutions that must be run for specific parameter values, excitations, and initial conditions.

The engineer must not forget that the model being analyzed is only an approximate mathematical description of the system, rather than the physical system itself. Conclusions based on equations that required a number of assumptions and simplifications in their development may or may not apply to the actual system.

Unfortunately, the more faithful a model is in describing the actual system, the more difficult it is to obtain general results.

One procedure is to use a simple model for analytical results and design, and then to use a different model to verify the design by means of computer simulation. In every complex system, it may be feasible to incorporate actual hardware components into the simulation as they become available, thereby the corresponding parts of the mathematic model could be eliminated.

1.4 Principle of Automatic Control Systems

1.4.1 Control and Control Systems

Control systems exist in a virtually infinite variety, both in the type of application and the level of sophistication, and can be found in nearly every facet of our daily life, although we are on such intimate terms with these systems that we often take them very much for granted and tend not to be aware of them.

As listed in Table 1.1, our body temperatures are controlled with excellent accuracy by our cardiovascular systems. Another type of control system is operating when we fetch objects on desks. The objects can be obtained because our hands move by the exact instructions that come from our brain through observation. In biological systems, the control schemes are quite minimal because the control system has been superbly designed by the long process of evolution. The larger birds of prey, such as hawks and eagles, are a good example of this. These birds can hover nearly motionlessly in a headwind by making very slight adjustments in the feathers on the trailing

Table 1.1: Some common systems.

Control object	Controller	Natural control systems		Man-made control systems	
		Manually controlled	Automatically controlled	Manually controlled	Automatically controlled
Human body temperatures	Cardiovascular systems		×		
Fetching objects on desks	Human brain and arm		×		
Hawks hovering motionlessly in a head wind	Hawk's feathers		×		
Turning angle of car's front wheels	Car's steering system			×	
Wheel's speed	Steam engine governor				×

edges of their wings. These control systems are provided by nature. There are also numerous man-made control systems. For example, a power steering system of a car is designed and made to control the turning angle of the car's front wheels. Hence, the running direction is controlled. In the process industries, including refineries and chemical plants, temperatures and levels can be held by control systems to usually constant values in the presence of various disturbances.

Both natural and man-made control systems share a common aim, which is to control or regulate a particular variable within certain operating limits, or varying according to certain discipline. A control system is, in the broadest sense, any interconnection of components providing a desired function. Information and constraints play the most significant roles, and meanwhile energy is necessary when a control system works. In fact, substance, energy, and information are the comprising classes of fundamental elements in any control system.

A system is said to be manually controlled, if the correcting action is performed by a human; otherwise, it is automatically controlled.

The arrangement that is used in control is called a controller, while the equipment or the process that is constrained or manipulated is called a controlled object or controlled plant. The combination of controller and controlled object is called a control system.

1.4.2 How Does an Automatic Control System Work?

How does a control system work? Let us consider two familiar examples.

The first one is about the constant temperature cabinet as shown in Fig. 1.2. The operator gets the temperature of the constant temperature cabinet from the thermometer. According to the thermometer, he gets the difference value and direction and makes a comparison with the desired temperature. In terms of the difference value and direction, the gentleman makes a control: if the actual temperature is higher

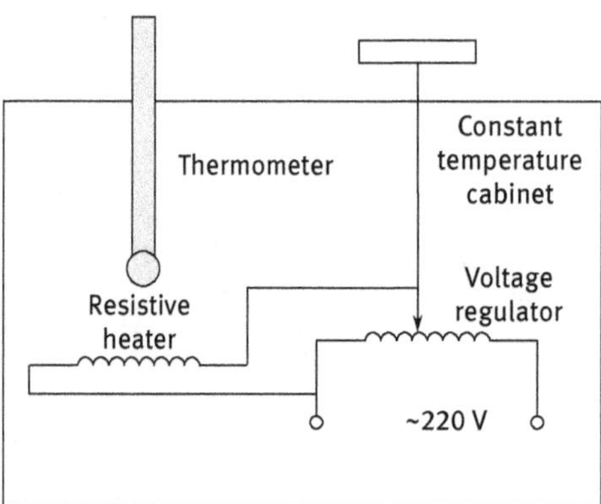

Fig. 1.2: The constant temperature cabinet.

than the desired, he will adjust the contactor of the voltage regulator to reduce the voltage. Thus, the temperature of the cabinet declines. Subsequently, the actual temperature reaches the desired level. Conversely, if the actual temperature is lower than the desired temperature, he will adjust the contactor of the voltage regulator to raise the voltage to get a higher temperature in the cabinet until the actual temperature equals the desired temperature. Finally, when the temperature in the cabinet reaches the desired level, the operator no longer needs to do anything to the contactor of the voltage regulator.

The second example is a historical example of the speed control of a steam engine. Prior to 1788, the speed of the steam engine was manually controlled. The worker read a speed on the indicator and increased or decreased the steam flow by setting steam flow valve. The process of manual control could be divided into two steps:

1. Observe the actual speed, and then compare the actual speed n_a and the desired speed n_d so as to attain the speed error $E = n_d \ n_a$.
2. Set steam flow valve according to the speed error E.
   ① If E is positive (which means the actual speed is lower than the desired), set the flow valve to increase the steam flow rate; the aim is to increase the actual speed.
   ② If E is negative (which means the actual speed is higher than the desired), the steam flow rate should be reduced.

In 1788, James Watt (1736–1819), who was a scientist and an inventor from England, introduced the famous governor that was an automatic control system, as shown in Fig. 1.3. The engine speed was monitored by the "lift" of rotating balls that worked against a spring. The initial tension of the spring was the reference, which was determined by the desired speed. As the rotating balls rose or fell, the steam control valve closed or opened via mechanical linkage in such a sense as to compensate for the speed change. Obviously, the automatic control process was similar to that of manual control. However, the difference in an automatic control process was that the observing, comparing, and setting (for the automatic control process) were performed by the controller automatically. Hence, an important conclusion that could be drawn is that the automatic control process, in fact, includes measuring errors of the controlled variables and correcting or diminishing them automatically (according to the errors).

According to the aforementioned principle analysis of the steam engine governor, the schematic diagram is illustrated in Fig. 1.4. The mechanical linkage plays roles of comparison and amplification, while the butterfly valve plays roles of amplification and control of steam flow rate. The steam engine is the controlled object; the output speed, also called the actual speed, is the controlled variable.

Furthermore, the schematic diagram for a general automatic control system could be obtained in a similar manner. The names of the components or variables in Fig. 1.4 should be merely substituted by the exact names of the other automatic

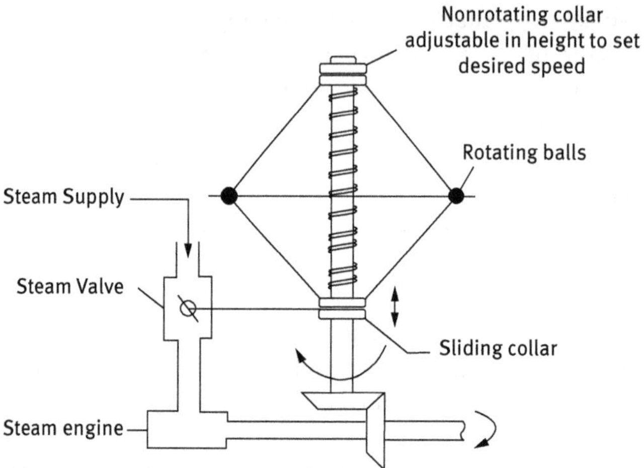

Fig. 1.3: A steam engine governor.

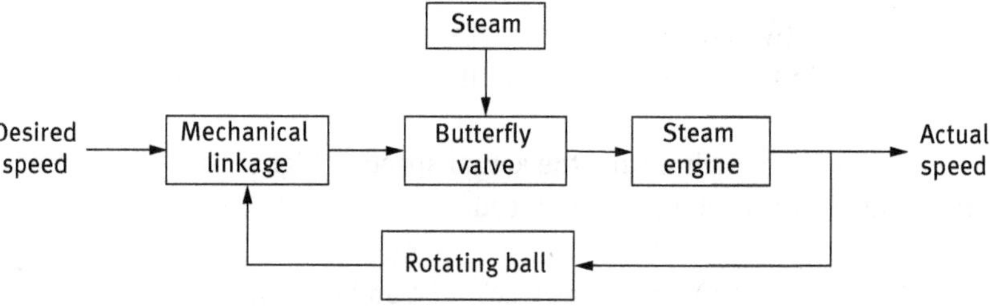

Fig. 1.4: A schematic diagram of the steam governor.

control systems. In the following section, we will give some explanation with details about the schematic diagram.

1.5 Structure of Control Systems

In this section, on the basis of the introduction for schematic diagram of the typical automatic control system, some useful and important terminologies will be given with the help of which the theories or principles could be easily and clearly understood.

1.5.1 Schematic Diagram for a Typical Automatic Control System

Generally, the schematic diagram for an automatic control system can be drawn as shown in Fig. 1.5. The signal "⊗" is called a summing point and is actually a comparator. The rectangular boxes are called blocks and are labeled with the titles

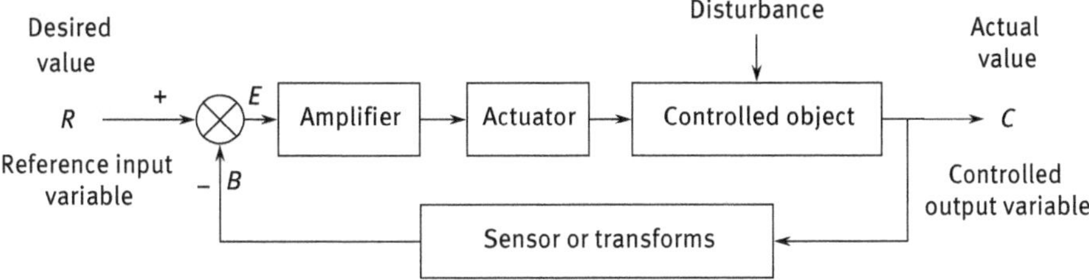

Fig. 1.5: A schematic diagram of a general automatic control system.

of the components. And the directed segments from a variable to a block represent the flow of the signals. In Section 1.5.2, we will explain the exact meaning for every block, including the relationship between two blocks, also known as the flow of signals. Here, we just need to focus on the summing point.

Designing a control system usually needs to meet the requirement of fulfilling a number of specifications (see Section 1.6), e.g., accuracy, speed of response, allowable overshoot, maximum duration of the settling time, and, of course, stability.

When designing a control system, however, we usually find that our first attempts are directed toward defining the structure of the control system correctly. If, however, the preliminary design for the control system proves to be unstable or close to instability, or if the system tends to be unstable when one is trying to improve the system response, or if any of the prescribed specification cannot meet the requirement of the system, one or more compensating devices must be added to the control system. In other words, compensators are needed in some cases to meet the requirements of the system or to improve the system's performance.

1.5.2 Terminologies

From Fig. 1.5, the following definitions can be extracted with details.

(1) A feedback (or closed-loop) control system is a control system that tends to maintain a prescribed relationship of one system variable to another by comparing the functions of these variables and using the difference as a means of control.

(2) The controlled output variable C is that quantity or condition of the controlled system that is directly measured and controlled.

(3) The reference input variable R is a signal established as a standard of comparison for a feedback control system. A constant value of R is also termed as a set point.

(4) The primary feedback signal B is a function of the controlled variable C, which is compared with the reference input R to obtain the actuating signal.

(5) The actuating signal E is equal to the reference input R minus primary feedback signal B.
(6) A disturbance U is a signal other than R, which tends to affect the value of the controlled variable C.
(7) The system error E is equal to the ideal or desired value R (also R is the reference input variable) minus the value of the primary feedback signal B in the absence of an indirectly controlled system.
(8) The feedback controller is a mechanism that measures the value of the controlled output variable C, accepts the value of the reference input variable R, and, as the result of a comparison, manipulates the controlled system in order to maintain an established relationship between the controlled variable and the command. The feedback controller comprises the feedback elements, control elements (including an amplifier and actuator), reference input variables, and summing point.
(9) The feedback elements, sensor/transformers, comprise the portion of the feedback control system that establishes the relationship between the primary feedback signal B and the controlled output variable C.
(10) The summing point is a descriptive symbol used in the block diagram to denote the algebraic summation of two or more signals. The direction of information flow is indicated by arrows.The algebraic nature of the summing point is represented by the plus or minus sign.

1.6 Characteristics of Control Systems

The characteristics of a control system will be discussed at great length in the following chapters. However, it is necessary to discuss the common characteristics for control systems, although different characteristics are needed for different control systems and different control tasks.

1.6.1 Stability

The steady-state output or response of a control system is usually a function of time. Mathematically, it varies with time, and may converge or not, as shown in Fig. 1.6.

If the difference between controlled output variable and reference input variable decays to zero, or the output converges, the system is said to be stable or the stability of the system is good (Fig. 1.6(a)). Otherwise, it is an unstable system or the stability of the system is bad (Fig. 1.6(b)). A practical closed-loop control system must be stable or must exist in a good stability.

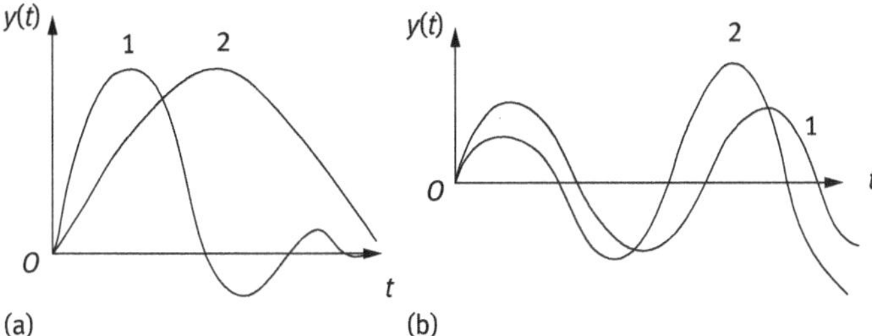

Fig. 1.6: The system stability: (a) stable system and (b) unstable system.

Open-loop control systems can never go unstable. However, when feedback is introduced, certain control systems can become unstable if care is not taken. Consequently, eliminating instability is the first concern of the control system designer.

1.6.2 Accuracy

The primary goal of feedback is to increase output accuracy by comparing the controlled output variable with a reference input variable. However, absolute accuracy cannot always be achieved, and some systems usually exhibit an error because of their simplicity. This error can usually be reduced to within acceptable limits, but penalties for this reduction are often incurred in the form of reduced stability.

These types of performance specifications are described only through steady-state errors in this book. In fact, dynamic errors are also significant when a control system is discussed. Numerous references are available related to this topic.

In Fig. 1.7, $x(t)$ and $y(t)$ are the input and the output signals for some systems, respectively. ε_{ss2} is the steady-state error for each system. In Fig. 1.7(a), the steady-state error ε_{ss2} goes to infinity. In Fig. 1.7(b), the steady-state error is equal to a constant R over K. In Fig. 1.7(c), the steady-state error is equal to zero. Above all, the accuracy of system represented by Fig. 1.7(c) is the best one among the three systems. About system accuracy, we will have a detailed explanation in Chapter 8.

1.6.3 Dynamic Properties

Dynamic properties indicate the transient response characteristics of control systems, usually related to step input. The transient response indices include overshoot percentage, peak time, settling time (according to the required accuracy), and rise time (Fig. 1.8).

The peak overshoot is expressed as a percentage of the desired step change. The settling time of a control system is the time taken by the output of the system to settle down within a tolerance band on either side of the desired value. The acceptable tolerance band δ must also be specified, and will depend upon the required accuracy

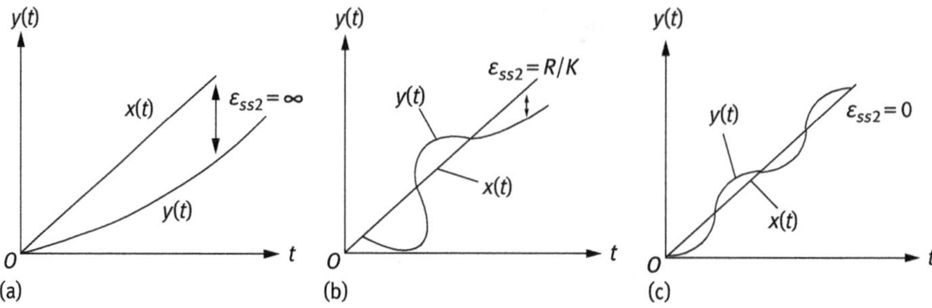

Fig. 1.7: The system accuracy (a) Steady state error goes to infinity (b) Steady state error goes to constant (c) Steady state error goes to zero.

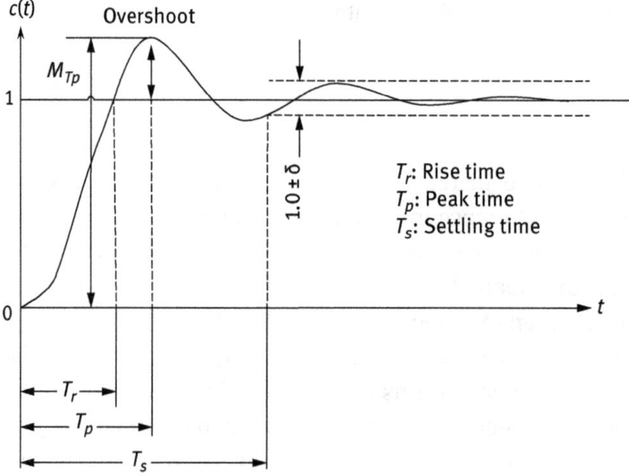

Fig. 1.8: Dynamic properties related to step input.

of the system. About ±5% or ±2% is reasonable for the step input. The rise time that is defined usually as the time when the response value increases from 5% to 95% of the final value in the initial stage of the output curve. Sometimes, it is also defined as the time taken by the system during which the response reaches a steady-state value. Peak time is defined as the time when the response curve first comes to the overshoot. Overshoot and peak time imply the smoothness of the system response, while the settling time and rise time show the system response speed.

1.6.4 Robustness

All systems are subjected to unwanted signals besides the primary reference input variable. If any signal, even though it carries intelligence, is unwanted, it is

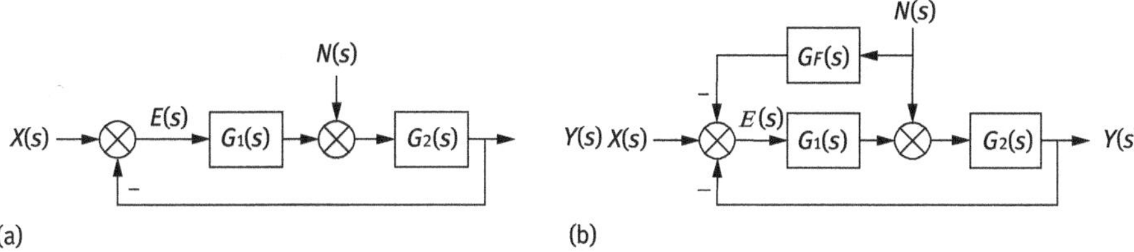

Fig. 1.9: The robustness for a system: (a) the original system with some disturbance signal $N(s)$ and (b) the feedback forward control.

considered to be noise. A feedback control process can be used to reduce the effects of noise.

It can be shown that the introduction of feedback drastically reduces the sensitivity of a control system to system disturbances and changes occurring in fixed component values (due to aging, etc.). The ability of a control system to reject various disturbances is the robustness of the control system.

The feedback control system depicted in Fig. 1.9(a) has some kind of disturbance signal $N(s)$, and the output will be influenced by this disturbance signal, especially for steady-state error of this system. A feedforward control link $G_F(s)$ can be introduced to reduce the steady-state error because of disturbance signal $N(s)$, as shown in Fig. 1.9(b). We can say the system in Fig. 1.9(b) is more robust than the system in Fig. 1.9(a). There are some explanations with details in the following chapters.

1.7 Classification of Control Systems (to Broadest Sense)

1.7.1 Open-Loop Control System

In an open-loop control system, the reference input variable r is fed directly to the controller at a value calculated to give the desired value c. Such a system is inaccurate, since any change in the status quo will affect the actual output. The difference between the input and output is due to two major effects: disturbances acting on the system and parameter variations of the system. For example, when you try to fetch an object in darkness, you cannot see anything. So, you may miss the object because of the disturbance, darkness, acting on you.

The toaster shown in Fig. 1.10 can be set for the desired darkness of the toasted bread. The setting of the "darkness" knob, or timer, represents the input quantity, and the degree of darkness and crispness of the toast is the output quantity. If the degree of darkness is not satisfactory because of the lack of feedback, this condition can in no way automatically alter the length of heating time. Since the output

Fig. 1.10: An automatic toaster.

quantity has no influence on the input quantity, there is no feedback in this system. The heater portion of the toaster represents the dynamic part of the system, and the time unit is the reference selector. In a broader context, the values of the parameters of the control system are seldom precisely known, and they may change greatly with operating conditions as well. So, the open-loop control systems will normally not yield high performances.

1.7.2 Closed-Loop Control System

For a closed-loop control system, the manipulated variable is a function of the actuating signal E, which is the difference between the reference input variable and the monitored or measured value B of the controlled variable. The output will affect the actuating signal. Therefore, the closed-loop control system can usually produce high performance and may be of high accuracy, but it would be unstable if they were not well designed. Thus, the sophistication level of a closed-loop control system is generally higher than the open-loop control system in order to keep the closed-loop control system stable.

The automatic elevator shown in Fig. 1.11 is a closed-loop control system, which is of great importance for multistory buildings. A person in the elevator presses the button corresponding to the desired floor, producing an actuating signal that indicates the desired floor and turns on the motor to raise or lower the elevator. As the elevator approaches the desired floor, the actuating signal decreases in value and, with the proper switching sequences, the elevator stops at the desired floor and the actuating signal becomes zero.

Take an actual example about this automatic elevator system. Assume that the elevator is on the 10th floor when the time is 10:15:00. At the same time, Tommy presses the button and gets in the elevator box, and then he presses the button for the 22nd floor. After5 s, the elevator arrives at the 15th floor and Zech

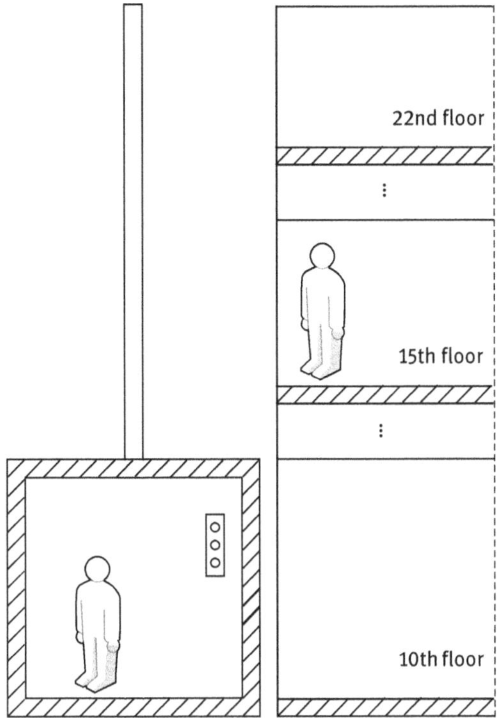

Fig. 1.11: The control system of an elevator.

Table 1.2: Given conditions for the elevator system.

	Now	Desired floor	Time at pressing button
Tommy	10th floor	22nd floor	10:15:00
Zech	15th floor	22nd floor	10:15:05
Elevator	10th floor	...	10:15:00
Elevator	15th floor	...	10:15:05

gets in the elevator. Zech also goes to the 22nd floor. All these given conditions are listed in Table 1.2. According to Fig. 1.11, some variables are also listed in Table 1.3.

In fact, at 10:15:00, 22nd floor should be the desired value and 10th floor should be the actual value. So, the primary feedback signal B should be 10th floor. Obviously, the error E should be equal to 12 floors (22 – 10 = 12). After 5 s, the actual value changes to 15th floor, and the error E also changes to 7 floors (22 – 15 = 7). Hence, on the basis of the above analysis, the disturbance variable is the 15th floor for this system.

Table 1.3: Variables for the elevator system.

Time	10:15:00	10:15:05
Desired value/reference input	22nd floor	22nd floor
Actual value/controlled output	10th floor	15th floor
B	10th floor	15th floor
E	12 floors	7th floors
Disturbance	...	Pressing signal in the 15th floor

1.8 Application of Control Theory in Mechanical Engineering Systems

It is believed that the first use of automatic control in Western civilization is dated back to the period of 300 BC. In modern time, the theory of control is applied in many fields.

(1) The theory of control is employed by economists, medical personnel, financial experts, political scientists, biologists, and engineers, to name but a few.
(2) Within the field of mechanical engineering, the heating system and water heater in a house or the heating, ventilation, and air conditioning system in a modern building, all employ automatic control systems.
(3) In aerospace, the control of aircrafts, helicopters, satellites, and missiles requires very sophisticated, advanced control systems.
(4) In ship-building industries, advanced control systems are often employed for steering and navigation.

1.9 Brief History of Automatic Control

It was as long ago as 1788 that James Watt found that a man controlling the opening and closing of steam valves was not the best way of keeping the speed of his steam engines constant. So, he developed the previously mentioned steam engine governor, which used the "lift" of rotating balls as a speed monitor to automatically shut off the steam as the speed tended to increase and vice versa. It was the first automatic control system brought to prominence, although undoubtedly not the first to be applied.

Instability, the unfortunate by-product of closed-loop control, was recognized early while working on steam engines and ship steering by James Clerk Maxwell, who published the first essay on cybernetics in 1868 and presented the concept of "feedback control." More lasting contributions, on the problems of stability in general, were made by Hurwitz (1875), Routh (1884), and

Lyapunov (1892) along with the essential works of mathematicians such as Laplace (1749–1827), Fourier (1758–1830), and Cauchy (1789–1857) based on which the modern engineering methods of analysis were found. Routh–Hurwitz stability criterion used algebraic methods merely to evaluate the stability of linear, time-invariant control system. However, Russian–Lyapunov's method can be employed to any control system.

The increase in the use of automatic control was understandably slow, since engineering was in its infancy. However, with the development of electronics, the application and theoretical understanding of closed-loop control system increased rapidly, since the feedback amplifier soon became a necessity. Above all, the war effort definitely accelerated the rate of advancement in science and technology in general, and control theory was no exception. The Second World War was the main stimulus to the progress of the control theory. Radar-controlled antiaircraft systems had to be designed. The large guns on ships needed to be stabilized against the roll of the ship if they were to fire accurately. Tracking radars needed control systems to help them maintain a "lock" on the target. Surface radar needed to be designed and controlled so that ships could navigate and fight in fog.

A most significant step was the work of Nyquist in 1932 with respect to stability in terms of the open-loop frequency response. Also of significance were the contributions of the pioneers of telecommunication, particularly Heaviside in the 1920s, leading to a better understanding of Laplace and Fourier transform. Thus, in 1934, Hazen presented the first precise analytical approach to the design of closed-loop control systems, drawing on the work of the aforementioned originators. From then on, accelerated by the Second World War, the need for faster and more accurate systems led to rapid development in the field, with the engineer gradually taking over the development from the mathematician. Prominent in this upsurge were Bode and Nichols for the frequency response techniques, Guillemin for the network synthesis approach, Evans for the root locus method, Weiner and Phillips for the statistical approach, and Tustin, Kochenburger, Lur'e, Ragazzini, Zadeh, Shannon, Bellman, Kalman, etc. for work on nonlinear, discrete, and multivariable systems. Most of the development may appear to have taken place in the USA, but much has also been done in parallel in the former USSR. Relatively little has stemmed from Europe.

The advances in control theory made during the war and in the following years are still with us.

It is worthwhile to mention one of the pioneers in the engineering cybernetics field, Qian Xuesen, who wrote a book *Engineering Cybernetics* (in English in the USA) in the early 1950s. The book was successively interpreted into Chinese, Russian, German, and French and was published. His contribution has greatly promoted the development of the engineering control theory.

Modern progress stems mainly from flight and process control and, rather unfortunately, from purely academic exercises, with particular accent on nonlinearity, optimization, and self-organizing systems. The state space approach is also receiving a lot of attention as an alternative method of analysis to the s-plane and frequency response methods used in this book.

In recent decades, modern control theories have developed rapidly. Many large-scale systems are being used to control such arrangements as in space shuttles, nuclear power stations, etc.

1.10 Organization of Book

This book, Fundamentals of Control Engineering, is designed to guide the students to achieve the following.

(1) Familiarity with the main concepts related to structures, characteristics, types, and principles of the control systems, as well as the system analysis and system modeling.

(2) Confidence in using Laplace transform method, including Laplace transform properties and Laplace transform inversion.

(3) Capability of formulating and understanding dynamic behavior of translational mechanical systems and RLC passive electrical systems.

(4) Familiarity with the transfer function of linear, time-invariant systems, including the transfer function of typical links, function block diagrams, signal flow diagrams, and Mason's gain formula.

(5) Understanding response analysis for the control system in time domain, especially performance specifications of first- or second-order system.

(6) Confidence in response analysis for control system in frequency domain, including Nyquist diagram and Bode diagram descriptions, minimum phase system, open-loop Bode diagram, and Nyquist stability criterion.

(7) Thorough acquaintance with stability analysis of the control system by using Routh–Hurwitz stability criterion.

(8) Appreciation of error analysis and calculation for the control system, and also familiarity with the main methods for reducing steady-state error of the control system.

1.11 Problems

P1.1. Give at least two examples from your daily life to illustrate open- and closed-loop control systems (at least 300 words).

P1.2. Make a schematic block diagram for the system shown in Fig. 1.10.

P1.3. Make a schematic block diagram for the system shown in Fig. 1.11.

P1.4. Explain the principle of water level control system in Fig. 1.12 (at least 300 words).

P1.5. Make a schematic block diagram for the system shown in Fig. 1.12 and point out its reference input variable, controlled variable, and disturbance variable.

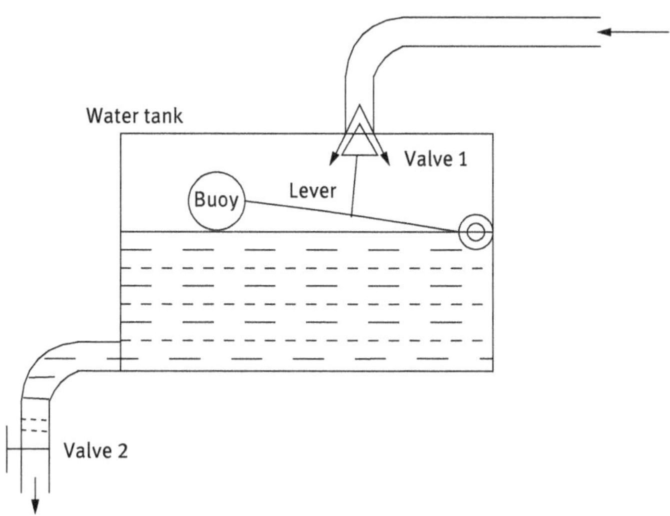

Fig. 1.12: A water level control system.

2 Laplace Transform Solution

Laplace transform is one of several powerful transformations that can be applied to analyze signals and engineering problems. A differential equation can be transformed into analgebraic equation with the help of the Laplace transform method to simplify the solving process. In the context of dynamic and control system analysis, Laplace transforms are applied to obtain the transfer functions, in turn, the block diagram representation and the solution of linear differential equations. Laplace transform has become one of the basic mathematical tools for analyzing the engineering control system.

2.1 Definition

Consider a function $f(t)$ defined for all $t \geq 0$; the Laplace transform $F(s)$ of $f(t)$ is defined as

$$F(s) = L[f(t)] = \int_0^{\infty} f(t)\mathrm{e}^{-st}\mathrm{d}t \tag{2.1}$$

where L stands for taking the Laplace transform. The symbol $L[f(t)]$ is read as "transform of," so that $L[f(t)]$ means "transform of $f(t)$." For the integral on the right-hand side (RHS) of eq. (2.1), the variable t vanishes after evaluation between the limits of integration. Thus, the resulting expression is a function of only s. Equation (2.1) defines the one-sided Laplace transform. There is a more general two-sided Laplace transform, which is useful in theoretical work, but is seldom used for solving for system responses.

In the definition of Laplace transform, its lower integral limitation is zero. If function $f(t)$ occurs at a jump $t = 0$, we need to figure out whether the lower integral limitation goes to zero from 0^+ or from 0^-. As a result, the Laplace transform is different for those two different integral limitations:

$$\begin{aligned} L_-[f(t)] &= \int_{0^-}^{\infty} f(t)\mathrm{e}^{-st}\mathrm{d}t \\ &= \int_{0^-}^{0^+} f(t)\mathrm{e}^{-st}\mathrm{d}t + \int_{0^+}^{\infty} f(t)\mathrm{e}^{-st}\mathrm{d}t \\ &= \int_{0^-}^{0^+} f(t)\mathrm{e}^{-st}\mathrm{d}t + L_+[f(t)] \end{aligned} \tag{2.2a}$$

$$L_+[f(t)] = \int_{0^+}^{\infty} f(t)\mathrm{e}^{-st}\mathrm{d}t \tag{2.2b}$$

https://doi.org/10.1515/9783110573275-002

2.2 Laplace Transforms of Common Functions

2.2.1 Step Function

Example 2.1
As shown in Fig. 2.1, find the Laplace transform for $f(t) = u(t)$, where $u(t)$ is the unit step function:

$$u(t) = \begin{cases} 0, & t < 0 \\ 1, & t \geq 0 \end{cases}$$

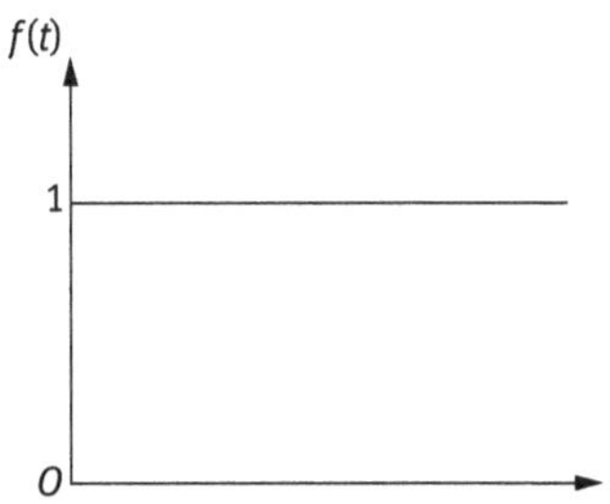

Fig. 2.1: The unit step function.

Solution
According to eq. (2.1),

$$\begin{aligned} F(s) &= \int_0^{\infty} f(t) \cdot \mathrm{e}^{-st} dt = \int_0^{\infty} 1 \cdot \mathrm{e}^{-st} dt \\ &= \int_0^{\infty} \left(-\frac{1}{s}\right) \cdot \mathrm{e}^{-st} d(-st) \\ &= -\frac{1}{s} \cdot \left[\mathrm{e}^{-st}\right]_0^{\infty} = -\frac{1}{s} \cdot (\mathrm{e}^{-s \cdot \infty} - \mathrm{e}^{0}) \\ &= \frac{1}{s} \end{aligned}$$

Note that the variable t has disappeared in the integration process and that the result is a function of s. If the amplitude of $u(t)$ is not unit, that is, $u(t)$ has a constant value R for all $t > 0$,

$$u(t) = \begin{cases} 0, & t < 0 \\ R, & t \geq 0 \end{cases}$$

We have

$$F(s) = \frac{R}{s}$$

2.2.2 Ramp Function

Example 2.2
As shown in Fig. 2.2, find the Laplace transform for $f(t) = t$, where $f(t)$ is the unit ramp function:

$$f(t) = \begin{cases} 0, & t < 0 \\ t, & t \geq 0 \end{cases}$$

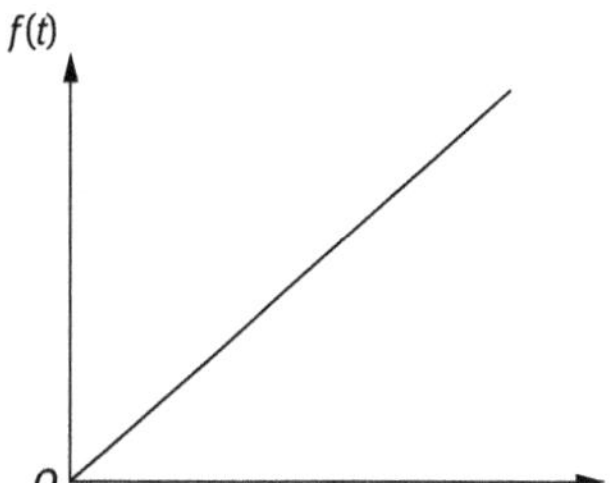

Fig. 2.2: The unit ramp function.

Solution

According to eq. (2.1),

$$F(s) = \int_0^{\infty} f(t) \cdot e^{-st} dt = \int_0^{\infty} t \cdot e^{-st} dt = \int_0^{\infty} -\frac{1}{s} \cdot te^{-st} d(-st) = -\frac{1}{s} \cdot \int_0^{\infty} td(e^{-st})$$

Using the integration by parts,

$$\int udv = uv - \int vdu$$

Let

$$\begin{cases} u = t \\ dv = de^{-st} \end{cases}$$

So,

$$\begin{cases} du = dt \\ v = e^{-st} \end{cases}$$

It can be shown that

$$F(s) = -\frac{1}{s} \cdot \left[te^{-st}\right]_0^{\infty} + \frac{1}{s} \cdot \int_0^{\infty} e^{-st} dt = 0 + \frac{1}{s} \cdot \left(-\frac{1}{s}\right) \cdot \left[e^{-st}\right]_0^{\infty} = \frac{1}{s^2}$$

If the amplitude of $f(t)$ is not unit,

$$f(t) = \begin{cases} 0, & t < 0 \\ At, & t \geq 0 \end{cases}$$

We have

$$F(s) = \frac{A}{s^2}$$

2.2.3 Pulse Function

Example 2.3
Find the Laplace transform for $\delta(t)$, where $\delta(t)$ is the unit pulse function, as shown in Fig. 2.3:

$$\delta(t) = \begin{cases} 0, & t \neq 0 \\ \infty, & t = 0 \end{cases}$$

The characteristics of the unit pulse function $\delta(t)$ are as follows, where $g(0)$ represents the initial value of the function $g(t)$ at the time $t = 0$:

$$\int_{-\infty}^{\infty} \delta(t)\mathrm{d}t = 1$$

$$\int_{-\infty}^{\infty} \delta(t)g(t)\mathrm{d}t = g(0)$$

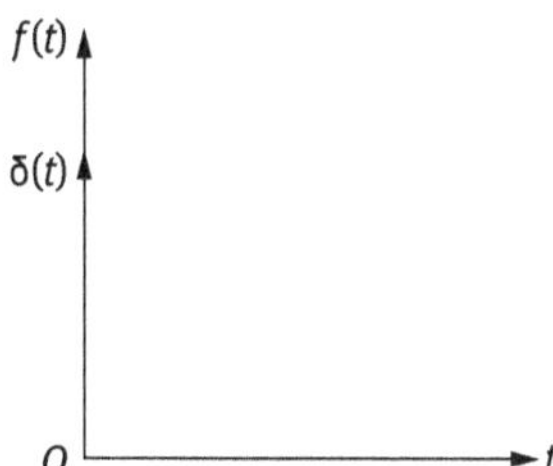

Fig. 2.3: The unit pulse function.

Solution
Because a jump occurs at the time $t = 0$ for the pulse function, $L_+[\delta(t)]$ cannot reflect all characteristics in the interval of $[0^-, 0^+]$. Considering eqs. (2.2a) and (2.2b), we obtain

$$\begin{aligned} F(s) &= \int_{0^-}^{\infty} f(t) \cdot \mathrm{e}^{-st}\mathrm{d}t = \int_{0^-}^{\infty} \delta(t) \cdot \mathrm{e}^{-st}\mathrm{d}t = \int_{0^-}^{0^+} \delta(t) \cdot \mathrm{e}^{-st}\mathrm{d}t + \int_{0^+}^{\infty} \delta(t) \cdot \mathrm{e}^{-st}\mathrm{d}t \\ &= \int_{0^-}^{0^+} \delta(t) \cdot g(t)\mathrm{d}t + \int_{0^+}^{\infty} 0 \cdot \mathrm{e}^{-st}\mathrm{d}t = g(0) = \mathrm{e}^{-st}|_{t=0} = 1 \end{aligned}$$

2.2.4 Exponential Function

Example 2.4
Find the Laplace transform for $f(t) = \mathrm{e}^{at}$, where $f(t)$ is the exponential function.

Solution

Depending on the value of the parameter a, the function $f(t) = e^{at}$ represents an exponentially decaying function ($a < 0$), a constant ($a = 0$), or an exponentially growing function ($a > 0$) for $t > 0$. According to eq. (2.1), in any case, the Laplace transform of the exponential function is

$$F(s) = \int_0^{\infty} f(t) \cdot e^{-st} dt = \int_0^{\infty} e^{at} \cdot e^{-st} dt = \int_0^{\infty} e^{-(s-a)t} dt = -\frac{1}{s-a} \cdot e^{-(s-a)t}\Big|_0^{\infty} = \frac{1}{s-a}$$

2.2.5 Trigonometric Function

Example 2.5a

Find the Laplace transform for sine function $f(t) = \sin\omega t$.

Solution

According to eq. (2.1),

$$F(s) = \int_0^{\infty} \sin \omega t \cdot e^{-st} dt$$

Relating to the Euler formula in the complex field,

$$\sin \omega t = \frac{1}{2j}\left(e^{j\omega t} - e^{-j\omega t}\right)$$

We obtain

$$\begin{aligned} F(s) &= \int_0^{\infty} \sin \omega t \cdot e^{-st} dt = \int_0^{\infty} \frac{e^{j\omega t} - e^{-j\omega t}}{2j} \cdot e^{-st} dt = \frac{1}{2j}\int_0^{\infty} \left[e^{-(s-j\omega)t} - e^{-(s+j\omega)t}\right] dt \\ &= \frac{1}{2j}\left[-\frac{e^{-(s-j\omega)t}}{s-j\omega} + \frac{e^{-(s+j\omega)t}}{s+j\omega}\right]_0^{\infty} = \frac{1}{2j}\left[\frac{1}{s-j\omega} - \frac{1}{s+j\omega}\right] = \frac{1}{2j} \cdot \frac{s+j\omega-(s-j\omega)}{(s-j\omega)(s+j\omega)} \\ &= \frac{1}{2j} \cdot \frac{2j\omega}{s^2+\omega^2} = \frac{\omega}{s^2+\omega^2} \end{aligned}$$

Example 2.5b

Find the Laplace transform for cosine function $f(t) = \cos\omega t$.

Solution

According to eq. (2.1),

$$F(s) = \int_0^{\infty} \cos \omega t \cdot e^{-st} dt$$

Relating to the Euler formula in the complex field,

$$\cos\omega t = \frac{1}{2}(e^{j\omega t} + e^{-j\omega t})$$

By the similar procedure of getting the Laplace transform for the sine function, we can show that

$$L[\cos\omega t] = \frac{1}{2}\int_0^{\infty}(e^{j\omega t} + e^{-j\omega t})e^{-st}dt = \frac{1}{2}\left\{\int_0^{\infty} e^{-(s-j\omega)t}dt - \int_0^{\infty} e^{-(s+j\omega)t}dt\right\}$$

$$= \frac{1}{2}\left\{-\frac{1}{s-j\omega}\left[e^{-(s-j\omega)t}\right]_0^{\infty} - \frac{1}{s+j\omega}\left[e^{-(s+j\omega)t}\right]_0^{\infty}\right\}$$

$$= \frac{1}{2}\left(\frac{1}{s-j\omega} - \frac{1}{s+j\omega}\right) = \frac{s}{s^2+\omega^2}$$

2.2.6 Power Function

Example 2.6
Find the Laplace transform for power function $f(t) = t^2/2$.

Solution
According to eq. (2.1),

$$F(s) = \int_0^{\infty}\frac{t^2}{2}e^{-st}dt = -\frac{1}{s}\int_0^{\infty}\frac{t^2}{2}de^{-st} = -\frac{t^2}{2s}e^{-st}\bigg|_0^{\infty} + \frac{1}{s}\int_0^{\infty} te^{-st}dt = \frac{1}{s}\int_0^{\infty} te^{-st}dt$$

$$= \frac{1}{s}\left(-\frac{t}{s}e^{-st}\bigg|_0^{\infty} + \frac{1}{s}\int_0^{\infty} e^{-st}dt\right) = -\frac{1}{s^3}\int e^{-st}d(-st) = -\frac{1}{s^3}e^{-st}\bigg|_0^{\infty} = \frac{1}{s^3}$$

2.2.7 Summary

Table 2.1 lists some useful Laplace transforms for common functions.

2.3 Laplace Transform Properties

The Laplace transform has several properties that are useful in finding the transforms of functions in terms of known transforms and in solving for the response of dynamic models. We will state, illustrate, and in most cases derive those properties that will be useful in later work.

Table 2.1: Laplace transforms of common functions.

Item	Original function $f(t)$	Image function $F(s)$
1	$\delta(t)$	1
2	$1(t)$	$\frac{1}{s}$
3	T	$\frac{1}{s^2}$
4	e^{-at}	$\frac{1}{s+a}$
5	te^{-at}	$\frac{1}{(s+a)^2}$
6	$\sin\omega t$	$\frac{\omega}{s^2+\omega^2}$
7	$\cos(\omega t)$	$\frac{s}{s^2+\omega^2}$
8	$t^n (n=1, 2, 3, \cdots)$	$\frac{n!}{s^{n+1}}$
9	$t^n e^{-at} (n=1, 2, 3, \cdots)$	$\frac{n!}{(s+a)^{n+1}}$
10	$\frac{1}{b-a}(e^{-at}-e^{-bt})$	$\frac{1}{(s+a)(s+b)}$
11	$\frac{1}{b-a}(be^{-bt}-ae^{-at})$	$\frac{s}{(s+a)(s+b)}$
12	$\frac{1}{ab}\left[1+\frac{1}{a-b}(be^{-at}-ae^{-bt})\right]$	$\frac{1}{s(s+a)(s+b)}$
13	$e^{-at}\sin\omega t$	$\frac{\omega}{(s+a)^2+\omega^2}$
14	$e^{-at}\cos\omega t$	$\frac{s+a}{(s+a)^2+\omega^2}$
15	$\frac{1}{a^2}(at-1+e^{-at})$	$\frac{1}{s^2(s+a)}$
16	$\frac{\omega_n}{\sqrt{1-\zeta^2}}e^{-\zeta\omega_n t}\sin\omega_n\sqrt{1-\zeta^2}t$	$\frac{\omega_n^2}{s^2+2\zeta\omega_n s+\omega_n^2}$
17	$\frac{-1}{\sqrt{1-\zeta^2}}e^{-\zeta\omega_n t}\sin\left(\omega_n\sqrt{1-\zeta^2}t-\varphi\right)$ $\varphi=\arctan\frac{\sqrt{1-\zeta^2}}{\zeta}$	$\frac{s}{s^2+2\zeta\omega_n s+\omega_n^2}$
18	$1-\frac{1}{\sqrt{1-\zeta^2}}e^{-\zeta\omega_n t}\sin\left(\omega_n\sqrt{1-\zeta^2}t+\varphi\right)$ $\varphi=\arctan\frac{\sqrt{1-\zeta^2}}{\zeta}$	$\frac{\omega_n^2}{s(s^2+2\zeta\omega_n s+\omega_n^2)}$

2.3.1 Multiplication by a Constant

To express $L[af(t)]$ in terms of $F(s)$, where a is a constant and $F(s) = L[f(t)]$, we use eq. (2.1) to write

$$L[af(t)] = \int_0^\infty af(t)\mathrm{e}^{-st}\mathrm{d}t = a\int_0^\infty f(t)\mathrm{e}^{-st}\mathrm{d}t = aF(s) \tag{2.3}$$

Thus, the Laplace transform of multiplying a function of time by a constant is equal to the Laplace transform of multiplying its transform by the same constant.

2.3.2 Superposition

The transform of the sum of the two functions $f(t)$ and $g(t)$ is

$$\begin{aligned} L[f(t)+g(t)] &= \int_0^\infty [f(t)+g(t)]\mathrm{e}^{-st}\mathrm{d}t \\ &= \int_0^\infty f(t)\mathrm{e}^{-st}\mathrm{d}t + \int_0^\infty g(t)\mathrm{e}^{-st}\mathrm{d}t \\ &= F(s)+G(s) \end{aligned} \tag{2.4}$$

Using eqs. (2.3) and (2.4), for any constant a and b and any transformable functions $f(t)$ and $g(t)$, we have the general superposition property as

$$L[af(t)+bg(t)] = aF(s)+bG(s) \tag{2.5}$$

As an illustration of the superposition property, we can evaluate $L[2 + 3\sin 4t]$ by using eq. (2.5) with eqs. (2.4) and (2.3),

$$L[2+3\sin 4t] = \frac{2}{s} + 3\left(\frac{4}{s^2+4^2}\right) = \frac{2s^2+12s+32}{s^3+16s}$$

2.3.3 Differential Theorem

Because we need to take the Laplace transform of each term in a differential equation when solving system models for their responses, we must derive expressions for the transform of derivatives in arbitrary order. Let $L[f(t)] = F(s)$, then the Laplace transform of nth-order derivatives of $f(t)$ can be

$$L\left[\frac{\mathrm{d}^n f(t)}{\mathrm{d}t^n}\right] = s^nF(s) - s^{n-1}f(0) - s^{n-2}f^{(1)}(0) - \cdots - sf^{(n-2)}(0) - f^{(n-1)}(0) \tag{2.6}$$

where $f(0), f^{(1)}(0), \ldots, f^{(n-1)}(0)$ are all the initial values for all-order derivatives of $f(t)$.

The common second-order differential theorem and first-order differential theorem are as follows:

$$L\left[\frac{\mathrm{d}^2 f(t)}{\mathrm{d}t^2}\right] = s^2F(s) - sf(0) - f^{(1)}(0)$$

$$L\left[\frac{\mathrm{d}f(t)}{\mathrm{d}t}\right] = sF(s) - f(0)$$

When the initial conditions are all equal to 0, that is, $f(0) = 0, f^{(1)}(0) = 0, \ldots, f^{(n-1)}(0) = 0$, then Laplace transform of the aforementioned equations are

nth-order derivative theorem,

$$L\left[\frac{\mathrm{d}^n f(t)}{\mathrm{d}t^n}\right] = s^n F(s) \tag{2.7a}$$

second-order derivative theorem,

$$L\left[\frac{\mathrm{d}^2 f(t)}{\mathrm{d}t^2}\right] = s^2 F(s) \tag{2.7b}$$

first-order derivative theorem,

$$L\left[\frac{\mathrm{d}f(t)}{\mathrm{d}t}\right] = sF(s) \tag{2.7c}$$

Example 2.7

Taking the Laplace transform of the following differential equation, where all the initial conditions (for nth-order derivatives) are zero:

$$5\frac{\mathrm{d}^3y(t)}{\mathrm{d}t^3} + 6\frac{\mathrm{d}^2y(t)}{\mathrm{d}t^2} + \frac{\mathrm{d}y(t)}{\mathrm{d}t} + 2y(t) = 4\frac{\mathrm{d}x(t)}{\mathrm{d}t} + x(t)$$

Solution

According to the constant-superposition property and differential theorem,

$$5s^3Y(s) + 6s^2Y(s) + sY(s) + 2Y(s) = 4sX(s) + X(s)$$

2.3.4 Integral Theorem

Let $L[f(t)] = F(s)$, then by taking the Laplace transform of nth-order integral of $f(t)$, we obtain

$$L\left[\int\cdots\int f(t)(\mathrm{d}t)^n\right] = \frac{1}{s^n}F(s) + \frac{1}{s^n}f^{(-1)}(0) + \frac{1}{s^{n-1}}f^{(-2)}(0) + \cdots + \frac{1}{s}f^{(-n)}(0) \tag{2.8}$$

where $f(0), f^{(-1)}(0), \ldots, f^{(-n)}(0)$ are all initial conditions of the $\int\ldots\int f(t)\mathrm{d}t$. The common second-order integral theorem and first-order integral theorem are as follows:

$$L\left[\int\int f(t)(\mathrm{d}t)^2\right] = \frac{1}{s^2}F(s) + \frac{1}{s^2}f^{(-1)}(0) + \frac{1}{s}f^{(-2)}(0)$$

$$L\left[\int f(t)\mathrm{d}t\right] = \frac{1}{s}F(s) + \frac{1}{s}f^{(-1)}(0)$$

When the initial conditions are all equal to 0, that is, $f(0) = 0, f^{(-1)}(0) = 0, \ldots, f^{(-n)}(0) = 0$, the Laplace transforms of the aforementioned equations are as follows for nth-order integral theorem, second-order integral theorem, and first-order integral theorem, respectively:

$$L\left[\int\cdots\int f(t)(\mathrm{d}t)^n\right] = \frac{1}{s^n}F(s) \tag{2.9a}$$

$$L\left[\int\int f(t)(\mathrm{d}t)^2\right] = \frac{1}{s^2}F(s) \tag{2.9b}$$

$$L\left[\int f(t)\mathrm{d}t\right] = \frac{1}{s}F(s) \tag{2.9c}$$

2.3.5 Initial Value Theorem

The initial and final value theorems are particularly useful in the analysis and design of control systems. More specifically, the initial value theorem can be used to confirm the initial situation of systems or components and can also be applied to the impulse response analysis. The final value theorem is frequently applied in the evaluation of steady-state errors of control systems.

Let $f(t)$ and its first-order derivative have Laplace transforms, so the initial value of the function $f(t)$ is

$$f(0) = \lim_{t \to 0} f(t) = \lim_{s \to \infty} s \cdot F(s) \tag{2.10}$$

2.3.6 Final Value Theorem

Let $f(t)$ and its first-order derivative have Laplace transforms, so the final value of the function $f(t)$ is

$$f(\infty) = \lim_{t \to \infty} f(t) = \lim_{s \to 0} s \cdot F(s) \tag{2.11}$$

Example 2.8
Evaluate the initial value of the function $f(t) = \mathrm{e}^{-\alpha t}$.

Solution
Taking the Laplace transform of $f(t)$,

$$F(s) = L\left[\mathrm{e}^{-\alpha t}\right] = \frac{1}{s+\alpha}$$

According to eq. (2.10),

$$f(0) = \lim_{t \to 0} f(t) = \lim_{s \to \infty} s \cdot F(s) = \lim_{s \to \infty} s \cdot \frac{1}{s+\alpha} = \lim_{s \to \infty} \frac{1}{1+\alpha/s} = 1$$

Example 2.9
Evaluate the final value of the function $f(t)$, and the Laplace transform of $f(t)$ is

$$F(s) = \frac{5}{s(s^2+s+2)}$$

Solution
According to eq. (2.11),

$$f(\infty) = \lim_{t \to \infty} f(t) = \lim_{s \to 0} s \cdot F(s) = \lim_{s \to 0} \frac{5}{s^2+s+2} = \frac{5}{2}$$

Here, it may be appropriate to note that the final value theorem is not valid if the denominator of $s \cdot F(s)$ contains any pole whose real part is zero or positive. An example is $F(s) = \frac{\omega}{s^2+\omega^2}$, which is the Laplace transform of $\sin\omega t$. In another aspect, if $\lim_{t \to \infty} f(t)$ does not exist as $t \to \infty$, one cannot use the final value theorem to obtain the steady-state value of a dynamic system or component.

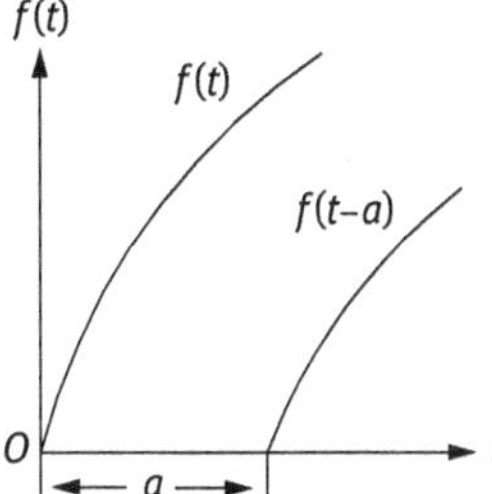

Fig. 2.4: Function *f*(*t*-*a*) and *f*(*t*).

2.3.7 Shifting Theorem in Time Domain (Delay Theorem)

Let $L\,[f(t)] = F\,(s)$, for any positive real number a,

$$L[f(t-a)] = \int_0^\infty f(t-a)\mathrm{e}^{-st}\mathrm{d}t = \mathrm{e}^{-as}F(s) \tag{2.12}$$

By means of Fig. 2.4, the function $f(t\text{-}a)$ is the delay function, where the delay time is a when compared with the function $f(t)$.

2.3.8 Shifting Theorem in Complex Domain

Let $L\,[f\,(t)] = F\,(s)$, for any constant a (real number or complex number),

$$L\left[\mathrm{e}^{-at}\cdot f(t)\right] = \int_0^\infty \mathrm{e}^{-at}\cdot f(t)\cdot \mathrm{e}^{-st}\mathrm{d}t = F(s+a) \tag{2.13}$$

Equation (2.13) states that multiplying a function $f(t)$ by e^{-at} is equivalent to replacing the variable s by the quantity $s + a$ wherever it occurs in $F(s)$. With the help of this property, we can easily derive several of the transforms from other items in Table 2.1,

$$L\left[\mathrm{e}^{-at}\cos\omega t\right] = \frac{s+a}{(s+a)^2+\omega^2}$$

$$L\left[\mathrm{e}^{-at}\sin\omega t\right] = \frac{\omega}{(s+a)^2+\omega^2}$$

$$L\left[t\mathrm{e}^{-at}\right] = \frac{1}{(s+a)^2}$$

2.3.9 Partial Fraction Method

In this section, the partial fraction method is applied to solve the following differential equation:

$$4\frac{d^2x}{dt^2} + \frac{dx}{dt} + 4x = 1 \tag{2.14a}$$

$$x(0) = \dot{x}(0) = 0 \tag{2.14b}$$

where the overdot denotes the derivative with respect to t. Taking the Laplace transform of eq. (2.14), one has

$$4\left[s^2X(s) - sx(0) - \dot{x}(0)\right] + sX(s) - x(0) + 4X(s) = \frac{1}{s}$$

Considering $x\ (0) = 0$ and $dx\ (0)/dt = 0$; therefore, the aforementioned equation becomes

$$4s^2X(s) + sX(s) + 4X(s) = \frac{1}{s}$$

$$[4s^2 + s + 4]X(s) = \frac{1}{s}$$

Therefore,

$$X(s) = \frac{1}{s(4s^2 + s + 4)}$$

By the partial fraction method, one can write the last equation as

$$\frac{1}{s(4s^2 + s + 4)} = \frac{A}{s} + \frac{Bs + C}{4s^2 + s + 4} \tag{2.15}$$

Making the numerators on both the sides of eq. (2.15) equal, one has

$$1 = A(4s^2 + s + 4) + s(Bs + C)$$

Expanding the RHS of the above equation, one has

$$1 = 4As^2 + As + 4A + Bs^2 + Cs$$

And arranging the terms according to the variable s, one obtains

$$1=(4A+B)s^2+(A+C)s+4A$$

Making the coefficients of both sides in the aforementioned equation to be equal according to the degree of variable s, one has

$$4A+B=0,\quad A+C=0,\quad 1=4A$$

Therefore,

$$A=\tfrac{1}{4},\quad C=-A=-\tfrac{1}{4},\quad B=-1$$

Substituting *A*, *B*, and *C* into the RHS of eq. (2.15), one obtains

$$\frac{1}{s(4s^2+s+4)}=\frac{1}{4}\left(\frac{1}{s}+\frac{-s-\frac{1}{4}}{s^2+\frac{1}{4}s+1}\right)$$

The terms inside the brackets on the RHS can be written as

$$X(s)=\frac{1}{4}\left[\frac{1}{s}-\frac{s+\frac{1}{8}}{\left(s+\frac{1}{8}\right)^2+\left(\frac{\sqrt{63}}{8}\right)^2}-\frac{\frac{1}{8}}{\left(s+\frac{1}{8}\right)^2+\left(\frac{\sqrt{63}}{8}\right)^2}\right]$$

$$=\frac{1}{4}\left[\frac{1}{s}-\frac{s+\frac{1}{8}}{\left(s+\frac{1}{8}\right)^2+\left(\frac{\sqrt{63}}{8}\right)^2}-\frac{\frac{\sqrt{63}}{8}\times\frac{1}{\sqrt{63}}}{\left(s+\frac{1}{8}\right)^2+\left(\frac{\sqrt{63}}{8}\right)^2}\right]$$

Finally, according to Table 2.1, one arrives at

$$x(t)=\frac{1}{4}\left[1-e^{-\frac{1}{8}t}\cos\frac{\sqrt{63}}{8}t-\frac{1}{\sqrt{63}}e^{-\frac{1}{8}t}\sin\frac{\sqrt{63}}{8}t\right].$$

In fact, one can get the above result according to the inverse Laplace transform, which is introduced in the following section.

2.4 Laplace Transform Inversion

When we use the Laplace transforms to solve the response of a system, first we need to find the transform $F(s)$ of a particular variable, such as the output. The final step in the process, known as transform inversion, is to determine the corresponding time function $f(t)$, where $f(t)=L^{-1}[F(s)]$, which reads as "$f(t)$ is the inverse transform of $F(s)$."

The process of finding the original function $f(t)$ corresponding to the image function $F(s)$ is called the Laplace transform inversion. The inverse of $F(s)$ or inverse Laplace transform of $F(s)$ is represented as

$$f(t) = L^{-1}[F(s)] = \frac{1}{2\pi \mathrm{j}} \int_{-\mathrm{j}\infty}^{+\mathrm{j}\infty} F(s)\mathrm{e}^{st}\mathrm{d}s$$

The simplified form is

$$f(t) = L^{-1}[F(s)] \tag{2.16}$$

The methods to solve the Laplace transform inversion are as follows:

(1) For simple image functions, we can find their original function $f(t)$ by Table 2.1, such as

$$f(t) = L^{-1}[F(s)] = L^{-1}\left[\frac{1}{s-a}\right] = \mathrm{e}^{at}$$

(2) For complicated image functions used in most engineering projects, we could first convert the complicated image functions into the summation of several simple image functions by means of partial fraction method. And then according to Table 2.1, one can find the original functions for every image function. At last, the sum of those original functions is the final result.

For example, $F(s)$ is the image function of the original function $f(t)$ and $F_n(s)$ is the image function of the original function $f_n(t)$. One can decompose $F(s)$ into the summation of several components,

$$F(s) = F_1(s) + F_2(s) + \cdots + F_{n-1}(s) + F_n(s)$$

According to Table 2.1, the original functions can be obtained as

$$\begin{aligned} f(t) &= L^{-1}[F(s)] = L^{-1}[F_1(s)] + L^{-1}[F_2(s)] + \cdots + L^{-1}[F_{n-1}(s)] + L^{-1}[F_n(s)] \\ &= f_1(t) + f_2(t) + \cdots + f_{n-1}(t) + f_n(t) \end{aligned}$$

Assume that we can write the transform $F(s)$ as the ratio of two polynomials $B(s)$ and $A(s)$, such that

$$F(s) = \frac{B(s)}{A(s)} = \frac{b_m s^m + b_{m-1}s^{m-1} + b_{m-2}s^{m-2} + \cdots + b_1 s + b_0}{a_n s^n + a_{n-1}s^{n-1} + a_{n-2}s^{n-2} + \cdots + a_1 s + a_0} \tag{2.17}$$

This function is called the rational function. A proper rational function is the one for which $m \leq n$, while a strictly proper rational function is the one for which $m < n$.

In order to write eq. (2.17) in the form of partial fractions, we need to rewrite $A(s)$ in factored form:

$$A(s) = (s - p_1)(s - p_2)(s - p_3) \cdots (s - p_n) \tag{2.18}$$

Note that the coefficient of the highest power of s in the denominator polynomial $A(s)$ has been assumed to be unity. If this is not the case for a given $F(s)$, we can always make the coefficient unity by dividing both $B(s)$ and $A(s)$ by a constant. The quantities $p_1, p_2, \ldots, p_i, \ldots, p_n$ are called the poles of the transform $F(s)$ and are those values of s for which $F(s)$ becomes infinite. The quantities p_i might be real numbers or complex numbers. Equation (2.17) can be written in the form of partial fraction,

$$F(s) = \frac{B(s)}{A(s)} = \frac{A_1}{s - p_1} + \frac{A_2}{s - p_2} + \cdots + \frac{A_n}{s - p_n} \tag{2.19}$$

The method of partial fraction expansion expresses a known transform $F(s)$ as the sum of less complicated transforms.

Using Table 2.1, we can identify the time functions that correspond to the individual transforms in the expansion and then use the superposition theorem to determine $f(t)$.

We consider that there are three different situations about the poles:

(1) All the poles of $F(s)$ are distinct and also real numbers.

(2) The procedure can be modified to include repeated poles, where two or more of the quantities $s_1, s_2, \ldots, s_n$ are equal.

(3) The poles are complex numbers.

2.4.1 Distinct Poles

The partial fraction expansion theorem states that if $F(s)$ is a strictly proper rational function with distinct poles, it can be written as

$$F(s) = \frac{B(s)}{A(s)} = \frac{B(s)}{(s - p_1)(s - p_2) \cdots (s - p_k) \cdots (s - p_n)}$$

$$= \frac{A_1}{s - p_1} + \frac{A_2}{s - p_2} + \cdots + \frac{A_n}{s - p_n} = \sum_{i=1}^{n} \frac{A_i}{s - p_i} \tag{2.20}$$

where $A_1, A_2, \ldots, A_n$ are constants.

From Table 2.1, the term $1/(s-p_i)$ is the transform of the time function $e^{p_i t}$. Then, according to the superposition formula given in eq. (2.5), it follows that for $t > 0$,

$$f(t) = A_1 e^{p_1 t} + A_2 e^{p_2 t} + \cdots + A_n e^{p_n t} = \sum_{i=1}^{n} A_i e^{p_i t} \tag{2.21}$$

We can find the n poles p_i by factorizing the denominator of $F(s)$ or, more generally, by finding the roots of $A(s) = 0$. We shall now develop a procedure for evaluating the n coefficients A_i, so that we can write $f(t)$ as a sum of exponential time functions by using eq. (2.21).

To get any coefficient A_k, we repeat eq. (2.20)

$$F(s) = \frac{B(s)}{A(s)} = \frac{B(s)}{(s-p_1)(s-p_2)\cdots(s-p_k)\cdots(s-p_n)}$$
$$= \frac{A_1}{s-p_1} + \cdots + \frac{A_k}{s-p_k} + \cdots + \frac{A_n}{s-p_n}$$

Then, multiplying both sides of eq. (2.20) by the term $(s-p_k)$ yields

$$\frac{A_1}{s-p_1}(s-p_k) + \cdots + \frac{A_k}{\cancel{s-p_k}}\cancel{(s-p_k)} + \cdots + \frac{A_n}{s-p_n}(s-p_k)$$
$$= \frac{B(s)}{(s-p_1)(s-p_2)\cdots\cancel{(s-p_k)}\cdots(s-p_n)} \cdot \cancel{(s-p_k)} = F(s)\cdot(s-p_k)$$

Furthermore,

$$\frac{A_1}{s-p_1}(s-p_k) + \cdots + A_k + \cdots + \frac{A_n}{s-p_n}(s-p_k)$$
$$= \frac{B(s)}{(s-p_1)(s-p_2)\cdots(s-p_{k-1})(s-p_{k+1})\cdots(s-p_n)} = F(s)\cdot(s-p_k)$$

Let $s = p_k$; the general expression for any coefficient is

$$A_k = F(s)\cdot(s-p_k)\big|_{s=p_k} \tag{2.22}$$

Example 2.10
Find Laplace transform inversion of

$$F(s) = \frac{(-s+5)}{(s+1)(s+4)}$$

Solution
Comparing $F(s)$ with eq. (2.20), we see that $B(s) = -s + 5$, so $m = 1$, and $A(s) = (s + 1)(s + 4)$, so $n = 2$. Because $A(s)$ is already in factored form, obviously, the poles are $s_1 = -1$ and $s_2 = -4$. Thus, we can rewrite the transform $F(s)$ in the form of eq. (2.20) as

$$F(s) = \frac{A_1}{s+1} + \frac{A_2}{s+4}$$

We find the coefficients of the partial fraction expansion to be

$$A_1 = \frac{(-s+5)}{(s+1)(s+4)} \cdot (s+1)\bigg|_{s=-1} = 2$$

$$A_2 = \frac{(-s+5)}{(s+1)(s+4)} \cdot (s+4)\bigg|_{s=-4} = -3$$

Hence, the partial fraction expansion of the transform is

$$F(s) = \frac{2}{s+1} - \frac{3}{s+4}$$

and, from Table 2.1, for $t > 0$, the time function $f(t)$ is

$$f(t) = 2e^{-t} - 3e^{-4t}$$

2.4.2 Repeated Poles

If the numbers of the repeated poles p_1 for $A(s) = 0$ are r, and if the remaining poles are still distinct, then $A(s)$ must be modified to be

$$A(s) = (s-p_1)^r (s-p_{r+1})(s-p_{r+2}) \cdots (s-p_n) \tag{2.23}$$

The partial fraction expansion of the $F(s)$ is)

$$F(s) = \frac{B(s)}{A(s)} = \frac{A_r}{(s-p_1)^r} + \frac{A_{r-1}}{(s-p_1)^{r-1}} + \cdots + \frac{A_1}{s-p_1} + \frac{B_{r+1}}{s-p_{r+1}} + \frac{B_{r+2}}{s-p_{r+2}} + \cdots + \frac{B_n}{s-p_n} \tag{2.24}$$

where, A_r, A_{r-1}, . . ., A_1 can be obtained by the series of the following equations:

$$\begin{aligned} A_r &= F(s)(s-p_1)^r\big|_{s=p_1} \\ A_{r-1} &= \left\{\frac{\mathrm{d}}{\mathrm{d}s}\left[F(s)(s-p_1)^r\right]\right\}_{s=p_1} \\ &\cdots \\ A_{r-j} &= \frac{1}{j!}\left\{\frac{\mathrm{d}^j}{\mathrm{d}s^j}\left[F(s)(s-p_1)^r\right]\right\}_{s=p_1} \\ &\cdots \end{aligned} \tag{2.25}$$

$$A_1 = \frac{1}{(r-1)!}\left\{\frac{\mathrm{d}^{r-1}}{\mathrm{d}s^{r-1}}[F(s)(s-p_1)^r]\right\}_{s=p_1}$$

The inverse transform of $\frac{1}{(s-p_1)^n}$ is

$$L^{-1}\left[\frac{1}{(s-p_1)^n}\right] = \frac{t^{n-1}}{(n-1)!}\mathrm{e}^{p_1 t} \tag{2.26}$$

For the coefficients of other real number poles like $B_{r+1}, B_{r+2}, \ldots, B_n$, we could use the method described in Section 2.4.1 to evaluate

$$B_k = [F(s)\cdot(s-p_k)]_{s=p_k}, \quad k = r+1, r+2, \cdots, n \tag{2.27}$$

Finally, we could get the Laplace transform inversionof $F(s)$ as

$$\begin{aligned} f(t) &= L^{-1}[F(s)] \\ &= \left[\frac{A_r}{(r-1)!}t^{r-1} + \frac{A_{r-1}}{(r-2)!}t^{r-2} + \cdots + A_2 t + A_1\right]\mathrm{e}^{p_1 t} \\ &\quad + B_{r+1}\mathrm{e}^{p_{r+1}t} + B_{r+2}\mathrm{e}^{p_{r+2}t} + \cdots + B_n\mathrm{e}^{p_n t} \end{aligned} \tag{2.28}$$

Example 2.11

Find Laplace transform inversion of

$$F(s) = \frac{5s+16}{(s+2)^2(s+5)}$$

Solution

The denominator of $F(s)$ is given in a factored form, so we note that the poles are $s_1 = s_2 = -2$ and $s_3 = -5$. Because of the repeated poles, the partial fraction expansion of $F(s)$ has the following form:

$$F(s) = \frac{A_{11}}{(s+2)^2} + \frac{A_{12}}{s+2} + \frac{A_3}{s+5}$$

Using eqs. (2.25) and (2.22) in order, we find that the coefficients are

$$A_{11} = (s+2)^2 F(s)\Big|_{s=-2} = \frac{5s+16}{s+5}\Big|_{s=-2} = 2$$

$$A_{12} = \left\{\frac{\mathrm{d}}{\mathrm{d}s}\left[\frac{5s+16}{s+5}\right]\right\}\Big|_{s=-2} = \frac{9}{(s+5)^2}\Big|_{s=-2} = 1$$

$$A_3 = (s+5)F(s)\Big|_{s=-5} = \frac{5s+16}{(s+2)^2}\Big|_{s=-5} = -1$$

Using the numerical values for the coefficients gives the partial fraction expansion of the transform

$$F(s)=\frac{2}{(s+2)^2}+\frac{1}{s+2}-\frac{1}{s+5}$$

For $t>0$, the corresponding time function is

$$f(t)=2te^{-2t}+e^{-2t}-e^{-5t}$$

2.4.3 Complex Poles

If p_1 and p_2 are a pair of conjugate complex number poles, the function $F(s)$ could be rewritten as

$$F(s)=\frac{b_m s^m+b_{m-1}s^{m-1}+b_{m-2}s^{m-2}+\cdots+b_1 s+b_0}{a_n s^n+a_{n-1}s^{n-1}+a_{n-2}s^{n-2}+\cdots+a_1 s+a_0}$$

$$=\frac{\alpha_1 s+\alpha_2}{(s-p_1)(s-p_2)}+\frac{A_3}{s-p_3}+\cdots+\frac{A_n}{s-p_n} \tag{2.29}$$

As described in Sections 2.4.1 and 2.4.2, to get the values of α_1 or α_2, we can multiply both sides of eq. (2.29) with $(s-p_1)(s-p_2)$ and let $s=p_1$ (or $s=p_2$),

$$\left[\frac{\alpha_1 s+\alpha_2}{(s-p_1)(s-p_2)}+\frac{A_3}{s-p_3}+\cdots+\frac{A_n}{s-p_n}\right]\cdot(s-p_1)(s-p_2)\bigg|_{s=p_1}=(\alpha_1 s+\alpha_2)_{s=p_1} \tag{2.30}$$

(1) Because p_1 is a complex number, both sides of the above equation comprise of complex numbers.
(2) Making two real parts of both sides of equation (2.30) correspondingly equal to get a new equation, and in the same way, two imaginary parts of both sides correspondingly equal to get another new equation.
(3) Using the aforementioned two equations, we can further get the values of α_1 and α_2.

Example 2.12
Find Laplace transform inversion of $F(s)=\frac{s+1}{s(s^2+s+1)}$.

Solution
Let $s^2+s+1=0$, and so two conjugate poles are $p_{12}=-0.5\pm j0.866$. Because of the conjugate poles, the partial fraction expansion of $F(s)$ has the form

$$F(s)=\frac{s+1}{s(s^2+s+1)}=\frac{\alpha_1 s+\alpha_2}{s^2+s+1}+\frac{A}{s}$$

Multiplying both sides with $(s-p_1)(s-p_2)$ and let $s=p_1$ (or $s=p_2$),

$$[F(s)\cdot(s-p_1)(s-p_2)]_{s=p_1}=\left[\frac{\alpha_1 s+\alpha_2}{s^2+s+1}+\frac{A}{s}\right]\cdot(s-p_1)(s-p_2)\Big|_{s=p_1}=(\alpha_1 s+\alpha_2)_{s=p_1}$$

$$\rightarrow\frac{s+1}{s(s^2+s+1)}\cdot(s^2+s+1)\Big|_{s=-0.5-j0.866}=(\alpha_1 s+\alpha_2)_{s=-0.5-j0.866}$$

$$\rightarrow\left(\frac{s+1}{s}\right)_{s=-0.5-j0.866}=(\alpha_1 s+\alpha_2)_{s=-0.5-j0.866}$$

$$\rightarrow\frac{0.5-j0.866}{-0.5-j0.866}=\alpha_1(-0.5-j0.866)+\alpha_2$$

Making two real parts of both sides of the equation correspondingly equal to get a new equation, and in the same way, two imaginary parts of both sides correspondingly equal to get another new equation.

$$\left.\begin{aligned}-0.5\alpha_1-0.5\alpha_2=0.5\rightarrow\alpha_1+\alpha_2=-1\\ 0.866\alpha_1-0.866\alpha_2=-0.866\rightarrow\alpha_1-\alpha_2=-1\end{aligned}\right\}\rightarrow\left\{\begin{aligned}\alpha_1&=-1\\ \alpha_2&=0\end{aligned}\right.$$

To confirm the coefficient A, referring to Section 2.4.1, we multiply both sides of the equation with s and also let $s=0$,

$$A=\left[\frac{s+1}{s(s^2+s+1)}\cdot s\right]_{s=0}=1$$

Substituting the above three coefficients, A, α_1, α_2, in the equation of $F(s)$, we get

$$F(s)=\frac{-s}{s^2+s+1}+\frac{1}{s}$$

And then, we apply transform to get some simple parts,

$$\begin{aligned}F(s)&=\frac{-s}{s^2+s+1}+\frac{1}{s}=\frac{1}{s}-\frac{s}{(s+0.5-0.866j)(s+0.5+0.866j)}\\ &=\frac{1}{s}-\frac{s}{(s+0.5)^2+0.866^2}=\frac{1}{s}-\frac{s+0.5-0.5}{(s+0.5)^2+0.866^2}\\ &=\frac{1}{s}-\frac{s+0.5}{(s+0.5)^2+0.866^2}+\frac{0.5}{(s+0.5)^2+0.866^2}\\ &=\frac{1}{s}-\frac{s+0.5}{(s+0.5)^2+0.866^2}+\frac{0.578\times0.866}{(s+0.5)^2+0.866^2}\end{aligned}$$

Finally, the Laplace transform inversion of $F(s)$ is

$$f(t) = L^{-1}[F(s)] = 1 - e^{-0.5t}\cos 0.866t + 0.578e^{-0.5t}\sin 0.866t, \quad t \geq 0$$

2.5 Problems

P2.1. Use the initial value theorem and the final value theorem to find $f(0+)$ and $f(\infty)$ when

$$F(s) = \frac{s^2 + 2s + 4}{s^3 + 3s^2 + 2s}$$

P2.2. Find the expression for $Y(s)/R(s)$ according to the differential equation (all initial conditions are zero)

$$\ddot{y}(t) + 3\dot{y}(t) + 6y(t) + 4\int y(t)\mathrm{d}t = 4r(t)$$

P2.3. Find the Laplace transform inversion of $F(s) = \dfrac{1}{s(s+1)^3(s+2)}$.

P2.4. Find the Laplace transform inversion of $F(s) = \dfrac{6(s+2)}{s(s^2+6s+12)}$.

3 Formulation and Dynamic Behavior of Translational Mechanical Systems

3.1 Introduction

3.1.1 Concepts of Mathematical Models

To analyze, synthesize, and design a dynamic system, an accurate mathematical model must be determined in advance. Mathematical models describe the motion patterns of physical systems. They quantitatively reveal the relationship between system's parameters and system's performance, and reveal the dynamic behavior of the system as well.

The derivation of the model is based on the fact that the dynamic system can be completely described by known differential equations or by experimental test data. The ability to analyze the system and to determine its performance depends on how well the characteristics can be expressed mathematically.

A good deal of the rest of the textbook is devoted to the analysis and design of control systems, on assumption that we already have an adequate model of the system.

3.1.2 Types of Mathematical Models

(1) Static models: those mathematical models that reveal the system motion patterns independent of time.

(2) Dynamic models: those mathematical models that reveal the system motion patterns dependent of time. Most mathematical models of control systems discussed in the book are dynamic, including the following:

   ① External models: they only reveal relationship between the input and the corresponding output of a control system, e.g., differential equation, transfer function, etc.

   ② Internal models: they reveal relationship among the input, output, and internal variables, e.g., state-space expression, etc.

A control system can also be described in frequency domain, i.e., frequency response. Bothanalytic expressions (e.g., differential equation, transfer function, time response, frequency response, state-space representation, etc.) and some diagrams (e.g., block diagram, signal flow diagram, Bode diagram, Nyquist diagram, etc.) can be employed to model a control system.

https://doi.org/10.1515/9783110573275-003

3.2 Variables

It is the first step that must be mastered by the control system engineer to write the differential equations for physical systems – electrical, mechanical, hydraulic systems, etc. The basic physical laws are given for the systems, the associated parameters are defined, and examples are included to show the application of the basic laws to physical equipment.

The symbols for the basic variables used to describe the dynamic behavior of translational mechanical systems are as follows:
(1) x, displacement in meters (m),
(2) v, velocity in meters per second (m/s),
(3) a, acceleration in meters per second squared (m/s^2),
(4) f, force in Newtons (N).

All these variables are functions of time. In general, however, we will add a t in parentheses immediately after the symbol only when it denotes an input or when we find doing so useful for clarity or emphasis.

Displacements are measured with respect to some reference condition, which is often the equilibrium position of the body or point in question. Velocities are normally expressed as the derivatives of the corresponding displacements. If the reference condition of a displacement is not indicated because it is not of interest, then the reference condition for the velocity needs to be given.

Two conventions used to define displacements are illustrated in Fig. 3.1 (a), (b). In Fig. 3.1(a), the variable x represents the displacement of the body from the fixed vertical wall, whereas in Fig. 3.1 (b), the reference or action position corresponding to $x = 0$ is not specifically shown.

Generally, the reference position will correspond to a condition of equilibrium for which the system inputs are constant and in which the net force on the body being considered is zero. Figure 3.1 (c), (d) indicate two methods of defining a velocity. All points on the body in Fig. 3.1 (c) must move with the same velocity, and so there is no possible ambiguity about which point has the velocity v. In Fig. 3.1 (d), the vertical line at the base of the arrow indicates that v is the velocity of the point labeled A. Forces can be represented by arrows pointing either into or away from a body, as depicted in Fig. 3.1 (e), (f), which are equivalent to one another.

Note that the arrows only indicate an assumed positive sense for the displacement, velocity, or force being considered, and by themselves do not imply anything about the actual direction of the motion or of the force at a given instant. If, for example, in Fig. 3.1 (e), (f), the force acting on the body is $f(t) = \sin t$, the force acts to the right for $0 < t < \pi$ and to the left for $\pi < t < 2\pi$, and it continues to change direction every π seconds. Note that an alternative way of describing the identical situation is to draw the arrow pointing to the left and then write $f(t) = -\sin t$. Reversing a reference

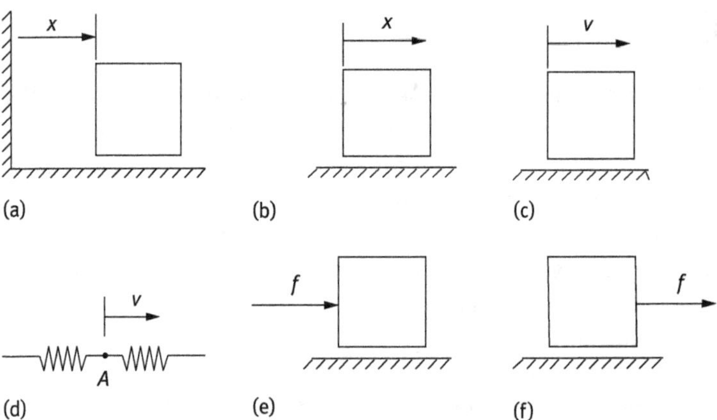

Fig. 3.1: Conventions for designating variables: (**a**) and (**b**) displacement; (**c**) and (**d**) velocity; and (**e**) and (**f**) force.

arrow is equivalent to reversing the sign of the algebraic expression associated with it. There is no unique way to choose reference directions on a diagram, but the equations must be consistent with whatever choice is made for the arrows.

Reference arrows for the displacement, velocity, and acceleration of a given point are invariably drawn in the same direction so that the equations

$$v = \frac{dx}{dt} \tag{3.1}$$

and

$$a = \frac{dv}{dt} = \frac{d^2x}{dt^2} \tag{3.2}$$

can be used. With this understanding, a reference arrow for acceleration is not shown explicitly on the diagrams, and for the same reason only the reference arrow for either the displacement or the velocity of a point (but not both) is shown in many examples.

3.3 Element Laws

Physical devices are represented by one or more idealized elements that obey laws involving the variables associated with the elements. As mentioned in Chapter 1, some degree of approximation is required in selecting the elements to represent a device, and the behavior of the combined elements may not correspond exactly to the behavior of the device. The major elements that we include in translational systems are mass, friction, and stiffness. The element laws for those three major elements relate the external force to the acceleration, velocity, or displacement associated with the element.

3.3.1 Mass

Figure 3.2(a) shows a mass M, which has units of kilograms (kg), subjected to a force f. Newton's second law states that the sum of the forces acting on a body is equal to the time rate of change of the momentum,

$$\frac{\mathrm{d}}{\mathrm{d}t}(Mv) = f \tag{3.3}$$

which, for a constant mass, can be written as

$$M\frac{\mathrm{d}v}{\mathrm{d}t} = f \tag{3.4}$$

For eqs. (3.3) and (3.4) to hold, the momentum and acceleration must be measured with respect to an inertial reference frame. For ordinary systems at or near the surface of the earth, the earth's surface is a very close approximation to an inertial reference frame, and so it is the one we use. The momentum, acceleration, and force are really vector quantities, but in this chapter the mass is constrained to move in a single direction, so we can write scalar equations.

We shall restrict our attention to constant masses and neglect relativistic effects so that we can use eq. (3.4). Hence, a mass can be modeled by an algebraic relationship between the acceleration $\mathrm{d}v/\mathrm{d}t$ and the external force f. For eq. (3.4) to hold, the positive senses of both $\mathrm{d}v/\mathrm{d}t$ and f must be the same, because the force will cause the velocity to increase in the direction in which the force is acting.

According to the relationship between the velocity $v(t)$ and the displacement $x(t)$, shown in Fig. 3.3(b), we can also get another expression with the second-order derivative of the $x(t)$,

$$f_M(t) = M\frac{\mathrm{d}}{\mathrm{d}t}v(t) = M\frac{\mathrm{d}^2}{\mathrm{d}t^2}x(t) \tag{3.5}$$

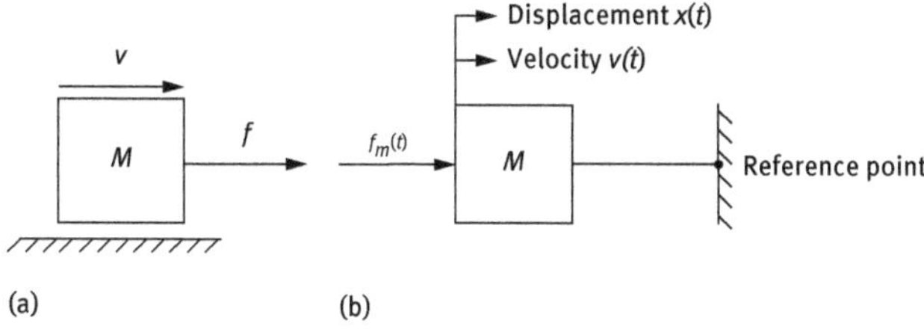

Fig. 3.2: The element laws for mass: **(a)** mass with force and **(b)** mass with major elements and reference point.

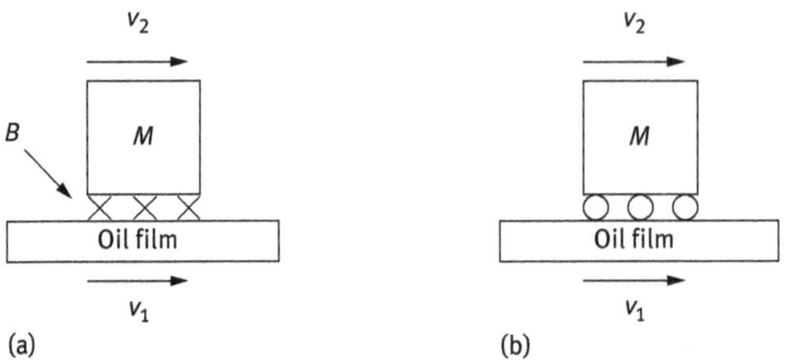

Fig. 3.3: The element laws for friction: (**a**) friction described by eq. (3.6) with $\Delta v = v_2 - v_1$ and (**b**) adjacent bodies with negligible friction.

3.3.2 Friction

Forces that are algebraic functions of the relative velocity between two bodies are modeled by friction elements. A mass sliding on an oil film that has laminar flow, as depicted in Fig. 3.3(a), is subject to viscous friction and obeys the linear relationship,

$$f = B\Delta v \tag{3.6}$$

where B has units of Newton-second per meter ($\mathrm{N \cdot s/m}$) and $\Delta v = v_2 - v_1$. The direction of a frictional force will be opposed to the motion of the mass. For eq. (3.6) to apply to Fig. 3.3(a), the force f exerted on the mass M by the oil film is to the left. (By Newton's third law, the mass exerts an equal force f to the right on the oil film.)

The friction coefficient B is proportional to the contact area and to the viscosity of the oil, and inversely proportional to the thickness of the film. A heavier mass would further compress the oil film, making it thinner and increasing the value of B.

Sometimes the frictional forces on adjacent bodies that have relative motion are small enough to be neglected. This might be the situation, for example, if the bodies are separated by bearings. The diagrams for such cases often show small wheels between the two bodies, as illustrated in Fig. 3.3(b), in order to emphasize the lack of frictional forces.

Viscous friction may also be used to model a dashpot, such as the shock absorbers on an automobile. As indicated in Fig. 3.4(a), a piston moves through an oil-filled cylinder, and there are small holes in the face of the piston through which the oil passes as the parts move relative to each other. The symbol shown in Fig. 3.4(b) is often used for a dashpot. Many dashpot devices involve a high rate of fluid through the orifices and have nonlinear characteristics. If the flow is laminar, then the element is again described by eq. (3.6). If the lower block in Fig. 3.3(a) or the cylinder of the dashpot in Fig. 3.4(a) is stationary, then $v_1 = 0$ and the element law reduces to $f = Bv_2$.

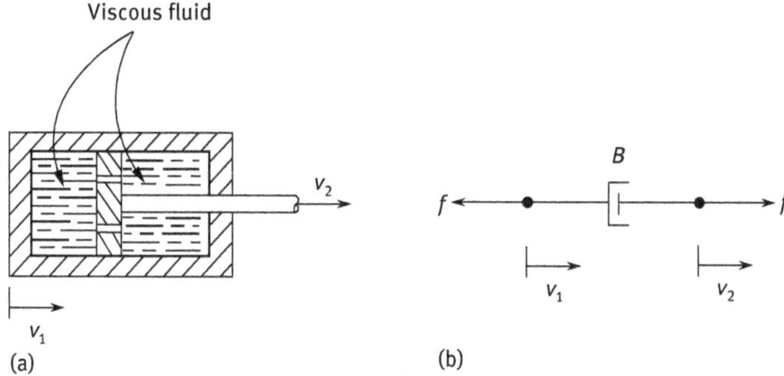

Fig. 3.4: The representation for the dashpot: (**a**) a dashpot and (**b**) its representation.

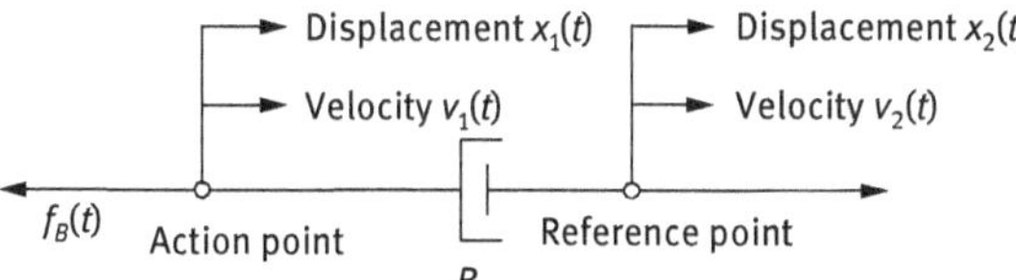

Fig. 3.5: Friction with the force.

If the dashpot or oil film is assumed to be massless and if the accelerations are to remain finite, then when a force f is applied to one side, a reaction force of equal magnitude must be exerted on the other side (either by a wall or by some other component) as shown in Fig. 3.4(b), again with $f = B(v_2-v_1)$. This means that in the system shown in Fig. 3.5, the force f is transmitted through the dashpot and exerted directly on the mass M.

According to the relationship between the velocity $v(t)$ and the displacement $x(t)$, we also can get another expression with the first-order derivative of the $x(t)$,

$$f_{\mathrm{B}}(t) = B\left[\frac{\mathrm{d}}{\mathrm{d}t}(x_1(t) - x_2(t))\right] = B\frac{\mathrm{d}\Delta x(t)}{\mathrm{d}t} = B[v_1(t) - v_2(t)] = B\Delta v(t) \tag{3.7}$$

3.3.3 Stiffness

Any mechanical element that undergoes a change in shape when subjected to a force can be characterized by a stiffness element, provided that only an algebraic relationship exists between the elongation and the force. The most common stiffness element is the spring, although most mechanical elements undergo some deflection when stressed. For the spring sketched in Fig. 3.6(a), d_0 is the length of the spring when no force is applied. x is the elongation caused by the force f. And $d(t)$ is the total length at any instant with the value of $d_0 + x$. The stiffness property

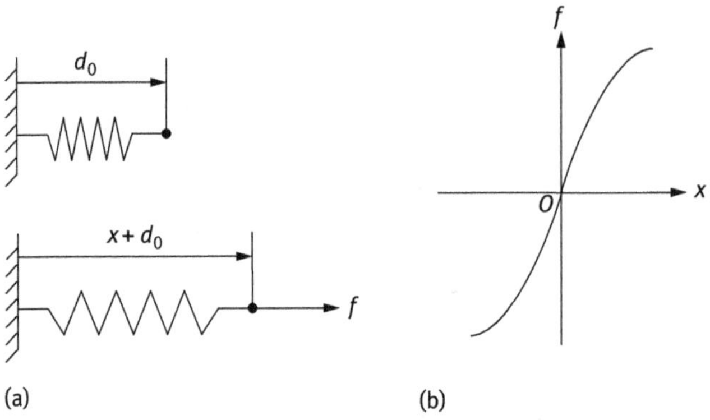

Fig. 3.6: Characteristic of a general spring: (**a**) the spring is sketched and (**b**) the curve about theelongation and the force.

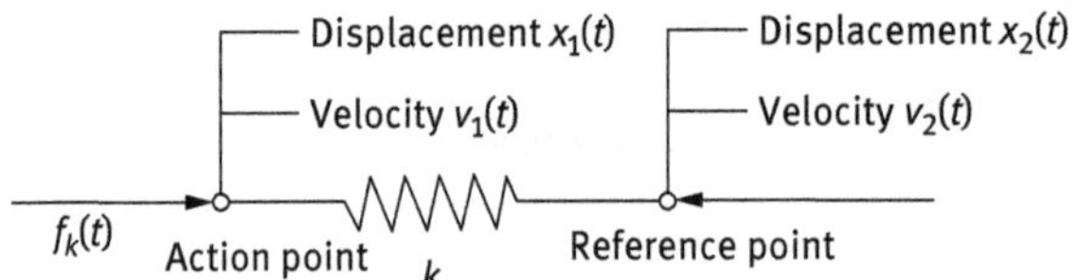

Fig. 3.7: Stiffness with the force.

refers to the algebraic relationship between x and f. For a linear spring, the curve in Fig. 3.6(b) is a straight line and $f = Kx$, where K is a constant with units of Newton's per meter (N/m).

When both sides of the spring have displacements, shown in Fig. 3.7, we have the expression depicting the relationship among the velocity, displacement, and force,

$$f_k(t) = k[x_1(t) - x_2(t)] = k\Delta x(t) = k\int[v_1(t) - v_2(t)]\mathrm{d}t = k\int \Delta v(t)\mathrm{d}t \tag{3.8}$$

3.4 Interconnection Laws

Having identified the individual elements in translational systems and having given equations to describe their behavior, we next present the laws that describe the manner in which the elements are interconnected. These include D'Alembert's law and the law of reaction forces.

3.4.1 D'Alembert's Law

D'Alembert's law is just a restatement of Newton's second law governing the rate of change of momentum. For a constant mass, we can write

$$\sum_i (f_{\text{ext}})_i = M\frac{\mathrm{d}v}{\mathrm{d}t} \tag{3.9}$$

where the summation over the index i includes all the external forces $(f_{\text{ext}})_i$ acting on the body. The forces and velocity are in general vector quantities, but they can be treated as scalars provided that the motion is constrained to be in a fixed direction. Rewriting eq. (3.9) as

$$\sum_i (f_{\text{ext}})_i - M\frac{\mathrm{d}v}{\mathrm{d}t} = 0 \tag{3.10}$$

This suggests that the mass in question can be considered to be in equilibrium, that is, the sum of the forces is zero, provided that the term $-M\cdot\mathrm{d}v/\mathrm{d}t$ is thought of as an additional force. This fictitious force is called the inertial force or D'Alembert's force, and including it along with the external forces allows us to write the force equation as one of equilibrium:

$$\sum_i f_i = 0 \tag{3.11}$$

This equation is known as D'Alembert's law. The minus sign associated with the inertial force in eq. (3.10) indicates that when $\mathrm{d}v/\mathrm{d}t > 0$, the force acts in the negative direction.

Many readers are more used to writing modeling equations using Newton's second law in eq. (3.9) rather than D'Alembert's law in eq. (3.11). The two are completely equivalent to one another. Frequently, the Newtonian formulation is used in beginning courses, and the D'Alembert procedure in later courses. However, it usually takes only a few minutes to become comfortable with the latter method. Using D'Alembert's law, where all forces including the inertial force are shown on the modeling diagrams, people are less likely to inadvertently leave out a term or have a sign error. Furthermore, this procedure produces equations that have the same form as those for other types of systems, including electrical and electromechanical ones. Those make it easier to talk about analogies.

In addition to applying eq. (3.11) to a mass, we can apply it to any point in the system, such as the junction between components. Because a junction is considered massless, the inertial force is zero in such a case.

3.4.2 The Law of Reaction Forces

To relate the forces exerted by the elements of friction and stiffness to the forces acting on a mass or junction point, we need Newton's third law regarding reaction

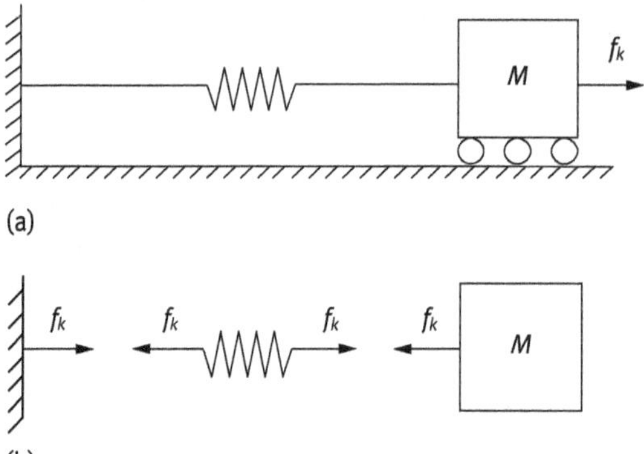

Fig. 3.8: Example of reaction forces: (**a**) the force exerted by the mass on the right end of the spring and (**b**) analyzing with Newton's third law.

forces. Accompanying any force of one element on another, there is a reaction force on the first element of equal magnitude and opposite direction.

In Fig. 3.8(a), for example, let f_k denote the force exerted by the mass on the right end of the spring, with the positive axis defined to be to the right. Newton's third law tells us that there acts on the mass a reaction force f_k of equal magnitude with its positive axis to the left, as indicated in Fig. 3.8(b). Likewise, at the left end, the fixed surface exerts a force f_k on the spring with the positive axis to the left, while the spring exerts an equal and opposite force on the surface.

3.5 Obtaining the System Model

The system model must incorporate both the element laws and the interconnection laws. The element laws involve displacements, velocities, and accelerations. Because the acceleration of a point is the derivative of the velocity, which in turn is the derivative of the displacement, we could write all the element laws in terms of x and its derivatives or in terms of x, v, and $\mathrm{d}v/\mathrm{d}t$. It is important to indicate the assumed positive directions for displacements, velocities, and accelerations. We shall always choose the assumed positive directions for a, v, and x to be the same, and so it will not be necessary to indicate all three positive directions on the diagram. Throughout this book, dots over the variables are used to denote derivatives with respect to time. For example $\dot{x} = \mathrm{d}x/\mathrm{d}t$, $\ddot{y} = \mathrm{d}^2y/\mathrm{d}t^2$.

3.5.1 Free-Body Diagrams

We normally need to apply D'Alembert's law, given by eq. (3.11), to each mass or junction point in the system that moves with a velocity that is unknown beforehand. To do so, it is useful to draw a free-body diagram for each such mass or point, showing all external forces and the inertial forces by arrows that define their positive senses. The element laws are used to express all forces except inputs in terms of displacements, velocities, and accelerations. We must be sure that the signs of these expressions are consistent with the directions of the reference arrows. After the free-body diagram is completed, we can apply eq. (3.11) by summing the forces indicated on the diagram, again taking into account their assumed positive axis. Normally all forces must be added as vectors, but in our examples the forces in the free-body diagram will be collinear and can be summed by scalar equations. The following two examples illustrate the procedure in some detail. The first system contains a single mass, and the second has two masses that can move with different velocities.

Example 3.1

Draw a free-body diagram and apply D'Alembert's law to write a modeling equation for the system shown in Fig. 3.9(a). The mass is assumed to move horizontally on frictionless bearings, and the spring and dashpot are linear. $f_a(t)$ is the applied force, and is also the input variable. $x(t)$ is the displacement of the mass, and is also the output variable. ($v(t)$ is just an intermediate variable and should not exist in the final differential equation.)

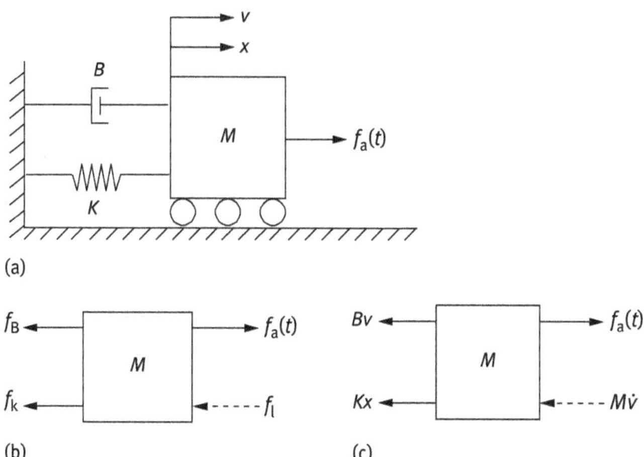

Fig. 3.9: The mass-damping–spring system: **(a)** translational system for Example 3.1; **(b)** free-body diagram; and **(c)** free-body diagram inclusing element laws.

Solution

The free-body diagram for the mass is shown in Fig. 3.9(b). The vertical forces on the mass (the weight $M{\cdot}g$ and the upward forces exerted by the frictionless bearings) have been omitted because these forces are perpendicular to the direction of motion. Hence, in our classical control theory, the vertical forces will not affect the motion with the horizontal direction.

The horizontal forces, which are included in the free-body diagram, are

(1) f_K, the force exerted by the spring, Kx.
(2) f_B, the force exerted by the dashpot, Bv.

(3) f_I, the inertial force/the fictitious force, $M \cdot dv/dt$.
(4) $f_a(t)$, the applied force, given.

The choice of directions for the arrows representing f_K, f_B, and f_I is arbitrary and does not affect the final result. However, the expressions for these individual forces must agree with the choice of arrows. The use of a dashed arrow is not an external force like the other three.

We next use the element laws to express the forces f_K, f_B, and f_I in terms of the element values K, B, and M and the system variables x and v. In Fig. 3.9(a), the positive direction of x and v is defined to be the right, and so the spring is stretched when x is positive, and compressed when x is negative. If the spring undergoes an elongation x, then there must be a tensile force Kx on the right end of the spring directed to the right and a reaction force $f_K = Kx$ on the mass directed to the left. In other words, if x is positive, the spring is stretched, and it therefore pulls mass M to the left, as seen in Fig. 3.9(c). As the arrow for f_K points to the left in Fig.3.9(b), we may relabel this force as Kx in Fig. 3.9(c). Note that if x is negative at some instant of time, the spring will be compressed and will exert a force to the right on the mass. Under these conditions, Kx will be negative and the free-body diagram will show a negative force on the mass to the left, which is equivalent to a positive force to the right. Although either of the result is the same, it is customary to assume that all displacements are in the assumed positive directions when determining the proper expressions for the forces.

Similarly, when the right end of the dashpot moves to the right with velocity v, a force $f_B = Bv$ is exerted on the mass to the left. Finally, because of eq. (3.10), the inertial force $f_1 = M\dot{v}$ must have its positive opposite to that of dv/dt. After trying a few examples, the reader should be able to draw a free-body diagram such as the one in Fig. 3.9(c) without first having to show explicitly the diagram in Fig. 3.9(b).

D'Alembert's law can now be applied to the free-body diagram in Fig. 3.9(c), with respect to the assumed arrow directions. If forces acting to the right are regarded as positive, the law yields

$$f_a(t) - (M\dot{v} + Bv + Kx) = 0$$

Replacing v by $\dot{x}$ and $\dot{v}$ by $\ddot{x}$, and rearranging the terms, we can rewrite this equation as

$$M\ddot{x} + B\dot{x} + Kx = f_a(t) \tag{3.12}$$

In some examples, the displacement variable associated with each object is expressed with respect to its own fixed reference position. However, the position of an object is sometimes measured with respect to some other moving body, rather than from a fixed reference. Such a relative displacement is used as one of the variables in the following example.

Example 3.2

Draw the free-body diagrams for the two-mass system shown in Fig. 3.10(a) and use D'Alembert's law to write the two differential equations that depict mass 1 and mass 2, respectively. $x_1(t)$ and $x_2(t)$ are the displacements of the mass 1 and mass 2, respectively. f_a (t) is the applied force. K_1, K_2, and B are coefficients for spring 1, spring 2, and dashpot, respectively.

Solution

Because there are two masses that can move with different unknown velocities, a separate free-body diagram should be drawn for each one. This is shown in Figs 3.9(b) and 3.10(c). In Fig. 3.10(b), the forces K_1x_1 and $M_1\ddot{x}_1$ are similar to those in Example 3.1. As indicated in our earlier discussion of displacements, the net elongation of the spring and dashpot connecting the two masses is (x_2-x_1).

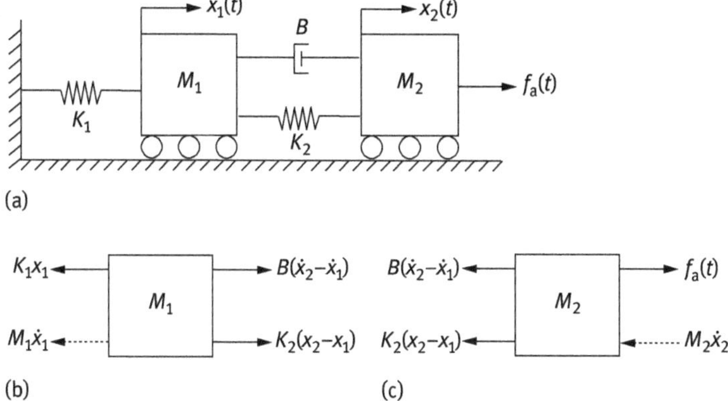

Fig. 3.10: The translational system and its free-body diagrams: (**a**) translational system for Example 3.2; (**b**) and (**c**) free-body diagrams.

Hence, a positive value of (x_2-x_1) results in a reaction force by the spring K_2 to the right on M_1 and to the left on M_2, as indicated in Fig. 3.10(b) and (c). Of course, the force on either free-body diagram could be labeled $K_2\,(x_1-x_2)$, provided that the corresponding reference arrow was reversed. For a positive value of $(\dot{x}_2-\dot{x}_1)$, the reaction force of the middle dashpot is to the right on M_1 and to the left on M_2. As always, the inertial forces $M_1\ddot{x}_1$ and $M_2\ddot{x}_2$ are opposite to the positive directions of the accelerations.

Summing the forces on each free-body diagram separately and taking into account the directions of the reference arrows give the following pair of differential equations:

$$B(\dot{x}_2-\dot{x}_1)+K_2(x_2-x_1)-M_1\ddot{x}_1-K_1x_1=0$$
$$f_a(t)-M_2\ddot{x}_2-B(\dot{x}_2-\dot{x}_1)-K_2(x_2-x_1)=0$$

Rearranging the equations, we have

$$M_1\ddot{x}_1+B\dot{x}_1+(K_1+K_2)x_1-B\dot{x}_2-K_2x_2=0 \tag{3.13a}$$

$$-B\dot{x}_1-K_2x_1+M_2\ddot{x}_2+B\dot{x}_2+K_2x_2=f_a(t) \tag{3.13b}$$

In the force eq. (3.13a) for the mass M_1 in the last example, note that all the terms involving the displacement x_1 and its derivatives have the same sign. Similarly, in eq. (3.13b) for the mass M_2, all the terms with x_2 and its derivatives have the same sign. This is generally true for systems where the only permanent energy sources are associated with the external inputs.

Suppose that D'Alembert's law is applied to any mass M_i and that terms involving corresponding variables are collected together. Then all the terms involving the displacement x_i of the mass M_i should be expected to have the same sign. If they do not, the engineer should suspect that an error has been made and should check the steps leading to that equation. In the simplified equation for M_i, no general statement can be made about the signs of terms involving displacements other than x_i and its derivatives, because the other signs depend on the reference directions used for the definition of the variables.

Inputs are variables that are specified functions of time. These functions are completely known and do not depend on the values of the system's components. Inputs for translational mechanical systems may be either forces or displacements. A displacement input exists when one part of a system is moved in a predetermined way. We assume that the mechanism that provides the specified displacement has a source of energy sufficient to carry out the motion regardless of any retarding forces that might come from the system components. Similarly, a force input is assumed to be available over whatever range of displacements may result.

Example 3.3

Draw a free-body diagram, including the effect of gravity, and find the differential equation describing the motion of the mass shown in Fig. 3.11(a). x is the displacement of the mass. $f_a(t)$ is the applied force. K and B are coefficients for spring and dashpot, respectively.

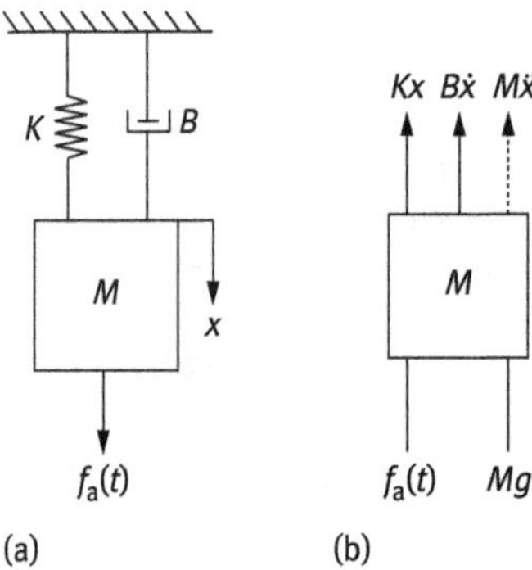

Fig. 3.11: The translational system with vertical motion and its free-body diagram: (**a**) translation system with vertical motion and (**b**) free-body diagram.

Solution

Assume that x is the displacement from the position corresponding to a spring that is neither stretched nor compressed. The gravitational force on the mass is Mg, and we include it in the free-body diagram shown in Fig. 3.11(b) because the mass moves vertically. By summing the forces on the free-body diagram, we obtain

$$M\ddot{x} + B\dot{x} + Kx = f_a(t) + Mg \tag{3.14}$$

3.5.2 Parallel Combinations

In some cases, two or more springs or dashpots can be replaced by a single equivalent element. Two springs or dashpots are said to be in parallel if the first end of each is attached to the same body and if the remaining ends are also attached to a common body. We shall consider a specific example before formulating a general rule.

Example 3.4
The system shown in Fig. 3.12(a) includes two linear springs between wall and mass.
(1) Write the differential equation describing the motion of the mass. (Assume that the springs have the same unstretched length.)
(2) Find the spring coefficient K_{eq} for a single spring that could replace K_1 and K_2. (Assume that the springs have the same unstretched length.)

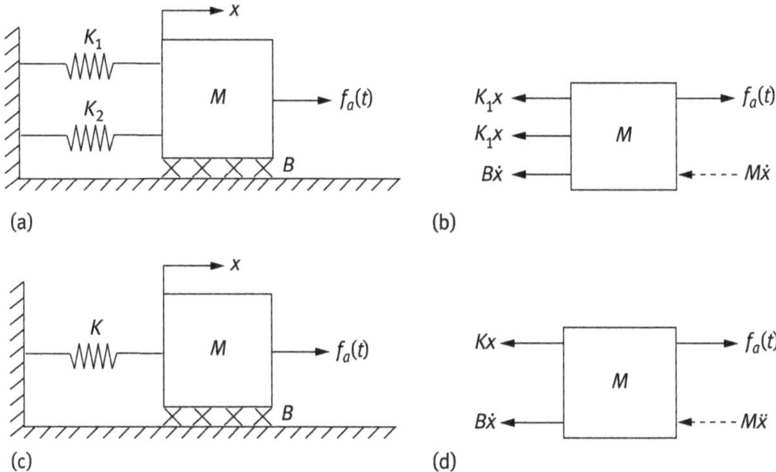

Fig. 3.12: The mass–spring system for Example 3.4: **(a)** translational system for Example 3.4; **(b)** free-body diagram when the springs have the same unstretched lengths; **(c)** a single equivalent sping K replaced by K_1 and K_2; and **(d)** free-body diagram when the combination of K_1 and K_2 *is replaced by a single equivalent spring.*

Solution (1)
If the unstretched lengths of the two springs are identical, then they will have the same elongation, denoted by x, when the mass is in motion. The free-body diagram shown in Fig. 3.12(b). Summing the forces gives

$$M\ddot{x} + B\dot{x} + (K_1 + K_2)x = f_a(t) \tag{3.15}$$

Solution (2)
If the combination of K_1 and K_2 is replaced by a single equivalent spring, then the system reduces to that shown in Fig. 3.12(c),

$$M\ddot{x} + B\dot{x} + Kx = f_a(t) \tag{3.16}$$

Comparing eqs. (3.15) and (3.16) reveals that

$$K_{eq} = K_1 + K_2 \tag{3.17}$$

Two parallel springs or dashpots have their respective ends joined, as shown in Fig. 3.13. From the last example, we see that for the parallel combination of two springs,

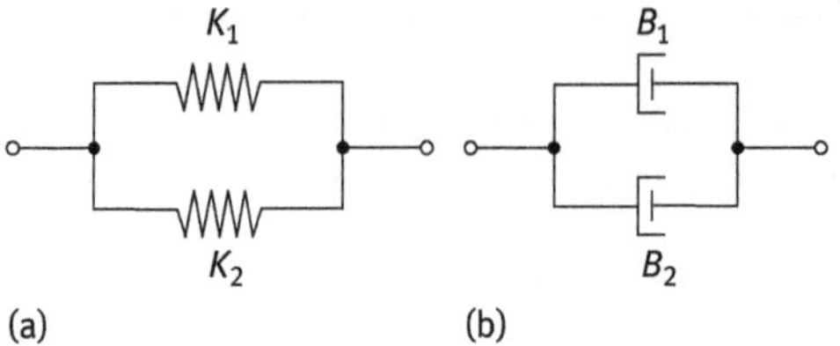

Fig. 3.13: Parallel combinations: (a) $K_{eq} = K_1 + K_2$ and (b) $B_{eq} = B_1 + B_2$.

$$K_{eq} = K_1 + K_2 \tag{3.18}$$

Similarly, it can be shown that for two dashpots in parallel, as in Fig. 3.13(b),

$$B_{eq} = B_1 + B_2 \tag{3.19}$$

The formulas for parallel stiffness or friction elements can be extended to situations that may seem slightly different from those in Fig. 3.13. The key requirement for parallel elements is that respective ends move with the same displacement. The individual ends need not be tied directly together in the figure depicting the system. Although one pair of ends may sometimes be connected to a fixed surface, in other cases both pairs of ends maybe free to move.

3.5.3 Series Combinations

Two springs or dashpots are said to be in series if they are joined at only one end of each element and if there is no other element connected to their common junction. The following example has a series combination of two springs and also illustrates the application of D'Alembert's law to a massless junction.

Example 3.5

When $x_1 = x_2 = 0$, the two springs shown in Fig. 3.14(a) are neither stretched nor compressed.

(1) Draw free-body diagrams for the mass M and for the massless junction A.

(2) Write the equations describing the system with the displacement x_1 not x_2. (Show that the motion of point A is not independent of that of the mass M and that x_1 and x_2 are proportional to one another.)

(3) Find K_{eq} for a single spring that could replace the combination of K_1 and K_2.

Solution (1)

The free-body diagrams are shown in Fig. 3.14(b), (c). Since there is no mass at point A, there is no inertial force in its free-body diagram. Summing the forces for each diagram gives

$$\begin{aligned} M\ddot{x}_1 + B\dot{x}_1 + K_1(x_1 - x_2) &= f_a(t) \\ K_2 x_2 &= K_1(x_1 - x_2) \end{aligned}$$

Solution (2)

Solving the second equation for x_2 in terms of x_1 gives

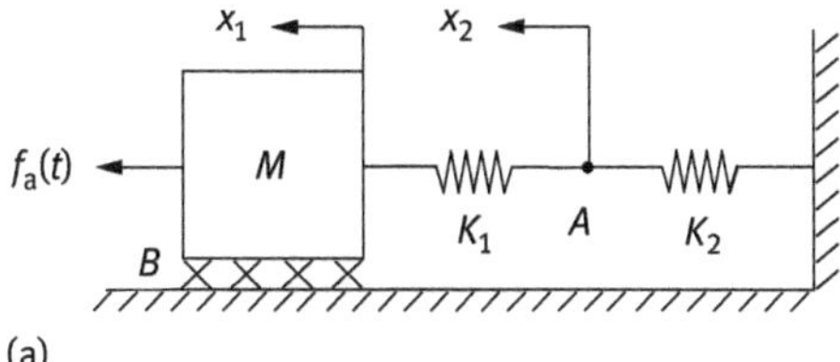

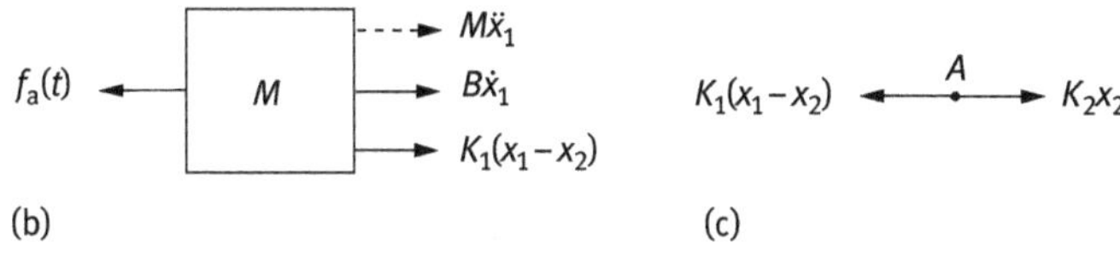

Fig. 3.14: The mass–spring system for Example 3.5: (**a**) translational system for Example 3.5; (**b**) free-body diagram for mass; and (c) free-body diagram for massless junction.

$$x_2 = \left(\frac{K_1}{K_1 + K_2}\right) x_1$$

which shows that the two displacements are proportional to one another. Substituting this expression back into the first equation, we have

$$M\ddot{x}_1 + B\dot{x}_1 + K_1\left[1 - \frac{K_1}{K_1 + K_2}\right] x_1 = f_a(t)$$

from which

$$M\ddot{x}_1 + B\dot{x}_1 + \frac{K_1 K_2}{K_1 + K_2} x_1 = f_a(t)$$

Solution (3)

This equation describes the system formed when the two springs in Fig. 3.14(a) are replaced by a single spring for which

$$K_{eq} = \frac{K_1 K_2}{K_1 + K_2} \tag{3.20}$$

Series combinations of stiffness and friction elements are shown in Fig. 3.15. It is assumed that no other element is connected to the common junctions. For the two springs in Fig. 3.15(a), the equivalent spring constant is given by eq. (3.20). For two dashpots in series, as in Fig. 3.15(b), it can be shown that

K_1 K_2 B_1 B_2

(a) (b)

Fig. 3.15: Series combinations: (a) $K_{eq} = K_1K_2/(K_1 + K_2)$ and (b) $B_{eq} = B_1B_2/(B_1 + B_2)$.

$$B_{eq} = \frac{B_1 B_2}{B_1 + B_2} \tag{3.21}$$

To reduce certain combinations of springs or dashpots to a single equivalent element, we may have to use the rules for both parallel and series combinations.

3.6 Summary

In this chapter, we have introduced the variables, element laws, and interconnection laws for linear, lumped-element translational mechanical systems. Our basic task in this chapter was to draw free-body diagrams and to write the corresponding equations describing the system. Either force or displacement inputs can be applied to any part of the system. An applied force is a known function of time, but the motion of the body to which it is applied is not known at the beginning of a problem. Conversely, a displacement input moves some part of the system with a known specified motion, but the force exerted by the external mechanism moving that part is normally not known.

Displacements may be measured with respect to fixed reference position or with respect to some other moving body. When relative displacement is used, it is important to keep in mind that the inertial force of a mass is always proportional to its absolute acceleration, not to its relative acceleration.

To model a system, we draw a free-body diagram and sum the forces for every mass or other junction point whose motion is unknown. The free-body diagram for a massless junction is drawn in the usual way, except that there is no inertial force. The modeling process can sometimes be simplified by replacing a series–parallel combination of stiffness or friction elements with a single equivalent element.

Special attention was given to linear systems that involve vertical motion. If displacements are measured from positions where the springs are neither stretched nor compressed, the gravitational forces must be included in the free-body diagrams for any masses that can move vertically. If, however, the displacements are measured with respect to the static equilibrium positions when the system is motionless and when no other external inputs are applied, then the gravitational forces do not appear in the final equation of motion.

3.7 Problems

P3.1. According to Fig. 3.16, the variables x_i and x_o is input displacement and output displacement, respectively. B is the damping coefficient. K_1 and K_2 are spring coefficients. Plot the free-body diagram for this system and find the differential equation.

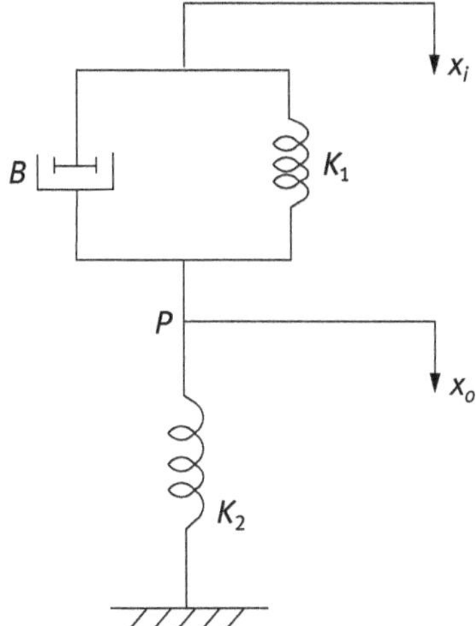

Fig. 3.16: The schematic diagram for P3.1.

P3.2. Fig. 3.17 shows a multiple elements mechanical translational system. The figure indicate the corresponding quantities.

Requirements:

(1) Illustrate the free-body diagrams of the masses.

(2) Write the differential equations describing the system.

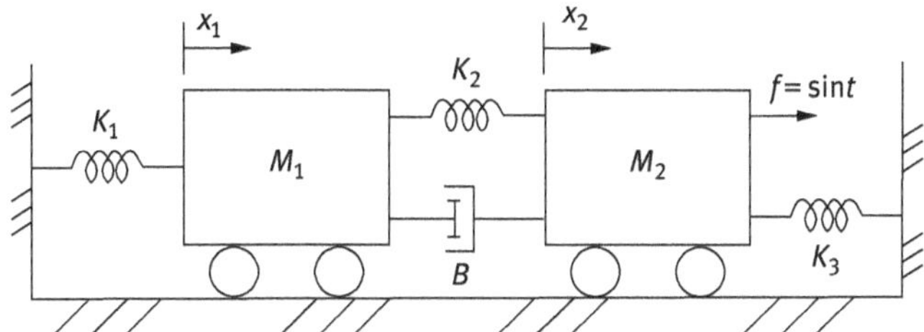

Fig. 3.17: A multiple elements mechanical translation system.

P3.3. In Fig.3.18, a force $f(t)$ is applied to the mass M_1. The sliding friction between the masses M_1, M_2, and the surface is indicated by the viscous friction coefficients μ_1

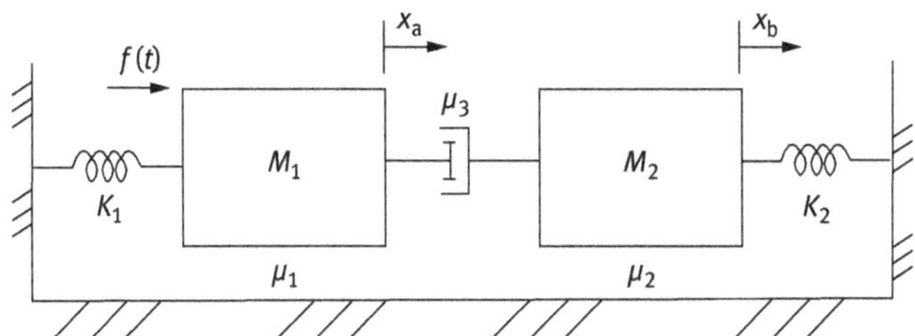

Fig. 3.18: A translational mechanical system composed of multiple elements.

and μ_2. The system equations can be written in terms of the two displacements x_a and x_b. For the two free bodies of the masses M_1 and M_2.
Requirements:
(1) Draw free-body diagrams for the masses M_1 and M_2.
(2) Write the differential equations describing the system.

P3.4. Find the differential equations for the system shown in Fig. 3.19. (x_i is input and x_o is output.)

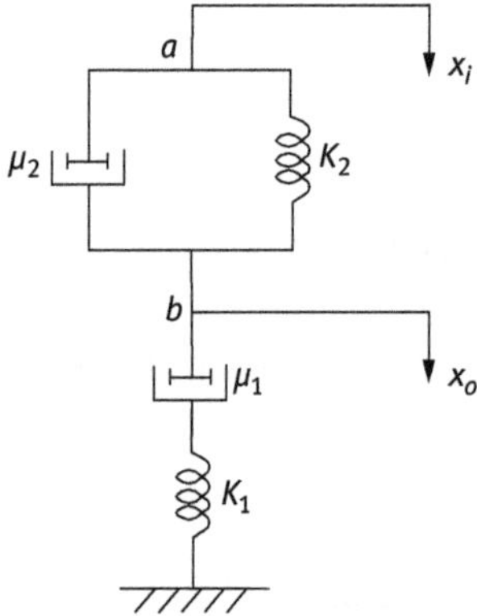

Fig. 3.19: The schematic diagram for P3.4.

4 Formulation and Dynamic Behavior of Electrical Systems

As expected at quite high frequencies, electrical circuits can usually be considered as an interconnection of lumped elements. In such cases involving a large and very important portion of the applications of electrical phenomena, we can model a circuit by using ordinary differential equations and we can apply the solution techniques discussed in this book.

In this chapter, we consider fixed linear circuits using the same approach that we have used for mechanical systems. We introduce the element and interconnection laws and then combine them to form procedures for finding the model of a circuit.

4.1 Element Laws

The elements in an electric circuit that we should consider are resistor, capacitor, inductor, and source. The first three of these are referred to as passive elements because, although they can store or dissipate energy that is present in the circuit, they cannot introduce additional energy. As shown in Fig. 4.1, they are analogous to the dashpot, mass, and spring for mechanical systems. In contrast, sources are active elements that can introduce energy into the circuit and that serve as the inputs. They are analogous to the force or displacement inputs for mechanical systems.

4.1.1 Resistor

A resistor is an element for which there is an algebraic relationship between the voltage across its terminals and the current through it, that is, an element that can be described by a curve of U_R versus i. A linear resistor is one for which the voltage and current are directly proportional to each other, that is, the voltage drop across a resistor is given by Ohm's law, which states that the voltage drop across a resistor equals the product of the current through the resistor and its resistance. Resistors absorb energy from the system. This voltage is written as

$$U_R = Ri \tag{4.1}$$

or

$$i = \frac{1}{R} U_R \tag{4.2}$$

https://doi.org/10.1515/9783110573275-004

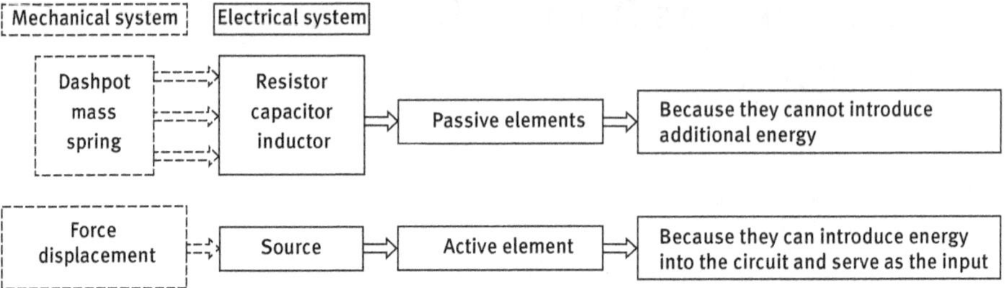

Fig. 4.1: The analogous relationship between a mechanical system and an electrical system.

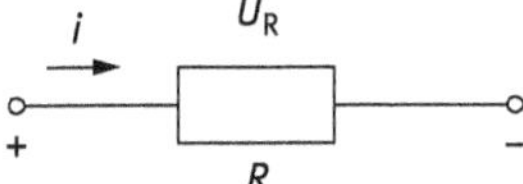

Fig. 4.2: A resistor and its variables.

where R is the resistance in ohms (Ω). A resistor and its current and voltage are denoted as shown in Fig. 4.2.

4.1.2 Capacitor

A capacitor is an element that obeys an algebraic relationship between the voltage and the charge, where the charge is the integral of the current. We use the symbol shown in Fig. 4.3 to represent a capacitor.

The positively directed voltage drop across a capacitor is defined as the ratio of the magnitude of the positive electric charge on its positive plate to the value of its capacitance. Its direction is from the positive plate to the negative one. The charge on a capacitor plate equals the time integral of the current entering the plate from the initial instant to the arbitrary time t, plus the initial value of the charge. For a linear capacitor, the capacitor voltage is written as follows:

$$U_C = \frac{1}{C}\int i \mathrm{d}t \tag{4.3}$$

where C is the capacitance in farads (F). For a fixed linear capacitor, the capacitance is a constant. If eq. (4.3) is differentiable, the element law for a fixed linear capacitor becomes

$$i = C\frac{\mathrm{d}U_C}{\mathrm{d}t} \tag{4.4}$$

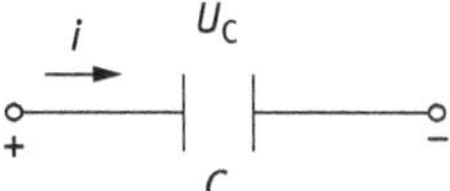

Fig. 4.3: A capacitor and its variables.

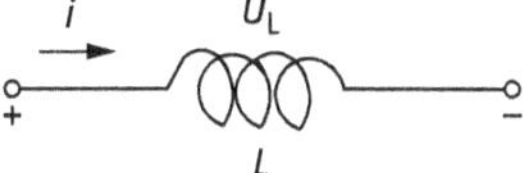

Fig. 4.4: An inductor and its variables.

4.1.3 Inductor

An inductor is an element for which there is an algebraic relationship between the voltage across its terminals and the derivative of the flux linkage. The symbol for an inductor and the convention for defining its current and voltage are shown in Fig. 4.4. For a linear inductor, the voltage drop across an inductor is given by Faraday's law, which is written as

$$U_{\mathrm{L}} = L\frac{\mathrm{d}i}{\mathrm{d}t} \tag{4.5}$$

where L is the inductance in henries (H). For a fixed linear inductor, L is constant.

4.2 Interconnection Laws

Two interconnection laws are used in conjunction with the appropriate element laws in modeling electrical circuits. These laws are known as Kirchhoff's voltage law and Kirchhoff's current law:

(1) The algebraic sum of voltages around a closed loop is equal to zero.
(2) The algebraic sum of currents flowing into a circuit node is equal to zero.

In other words, they can be restated as follows: in traversing any closed loop the sum of the voltage rises equals the sum of the voltage drops. The sum of the currents entering the junction equals the sum of the currents leaving the junction. Electrical system element models include resistors, capacitors, inductors, voltage sources, and current sources. The voltage sources are usually alternating-current or direct-current generators. The usual direct-current voltage source is a battery. The voltage drops appear across the three basic electrical elements: resistors, inductors, and capacitors. These elements have constant component values.

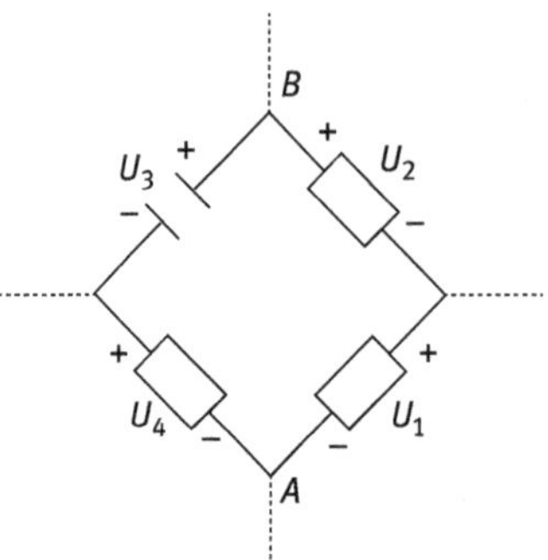

Fig. 4.5: Partial circuits to illustrate Kirchhoff's voltage law.

4.2.1 Kirchhoff's Voltage Law

When a closed path, that is, a loop is traced through every part of a circuit, the algebraic sum of the voltages across the elements that make up the loop must equal zero. This property is known as Kirchhoff's voltage law. For the voltages around any loop, it can be written as

$$\sum_j U_j = 0 \tag{4.6}$$

where U_j denotes the voltage across the jth element in the loop.

It follows that summing the voltages across individual elements in any two different paths from one point to another will give the same result. For instance, in the portion of a circuit sketched in Fig. 4.5, summing the voltages around the loop, going in a counterclockwise direction, and taking into account the polarities indicated on the diagram give

$$U_1 + U_2 - U_3 - U_4 = 0$$

Reversing the direction in which the loop is traversed yields

$$U_3 + U_4 - U_1 - U_2 = 0$$

Similarly, going from point A to point B by each of the two paths shown gives

$$U_1 + U_2 = U_3 + U_4$$

which is, of course, equivalent to both of the foregoing loop equations.

4.2.2 Kirchhoff's Current Law

When the terminals of two or more circuit elements are connected together, the common junction is referred to as a node. All the joined terminals are at the same

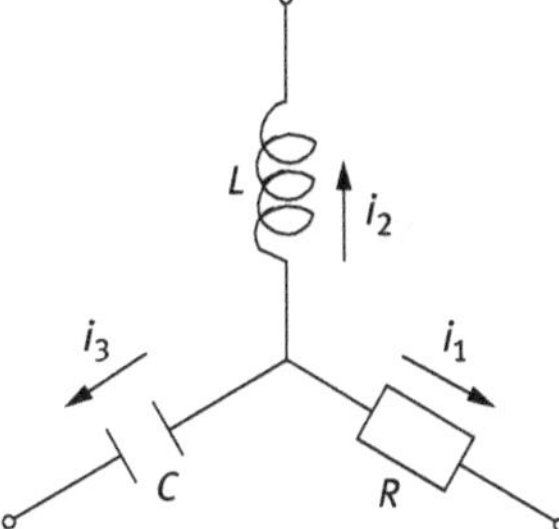

Fig. 4.6: Partial circuits to illustrate Kirchhoff's current law.

voltage and can be considered part of the node. Since it is not possible to accumulate every net charge at a node, the algebraic sum of the currents at any node must be zero at all times. This property is known as Kirchhoff's current law. For the currents at any node, it can be written as

$$\sum_j i_j = 0 \tag{4.7}$$

where the sum is over the currents through all elements joined to the node.

While applying eq. (4.7), we must take the directions of the current arrows into account. We should use a plus sign in eq. (4.7) for a current arrow directed away from the node being considered and a minus sign for a current arrow directed toward the node. This is consistent with the fact that the current i entering a node is equivalent to the current i leaving the node. For the partial circuit shown in Fig. 4.6, applying eq. (4.7) at the node to which the three elements are connected gives $i_1 + i_2 + i_3 = 0$. If we wish, we can also use Kirchhoff's current law in eq. (4.7) for any closed surface that surrounds part of the circuit.

4.2.3 The Nodal Method of Electrical Network Analysis

In the nodal method of electrical network analysis, one node in the network is usually chosen as the reference node and voltages between the reference node and other nodes are defined. Expressing the element currents in terms of node voltages and applying Kirchhoff's current law at each node except the reference node give the same number of independent simultaneous integro-differential equations as there are node voltages. If voltage sources are present, the unknowns include the currents through the voltage sources, while the sources fix node-to-node voltages. The rules for writing the integro-differential equations for each node are summarized as follows:

(1) The number of equations equals the number of unknown node voltages.
(2) One equation is written for each node.

(3) Each equation includes the following: ① The node voltage multiplied by the sum of all the admittances that are connected to this node. This term is negative. ② The node voltage at the other end of each branch multiplied by the admittance connected between the two nodes. This term is positive.

4.3 Analogue relationships among different systems

Dynamic characteristics of a control system can be described by a differential equation. To write the differential equations of the physical systems, the basic laws governing the performance of the basic elements of the systems are stated first and then applied to specific devices so as to obtain the differential equations of the performance. The differential equations for different physical systems may have the same form. The corresponding variables and parameters in two or more systems represented by equations of the same form are called analogs, and these systems are said to be similar if they have the differential equations of the same form. Physical similarity among different systems is listed in Tab. 4.1.

Figure 4.7 shows some analogue relationships between a mechanical system and an electrical system. We will give further explanation with details in the following chapter; here, we just use them to illustrate the analogue relation.

Fig. 4.7(a) gives a hydraulic integral link with the transfer function:

$$G(s) = \frac{X(s)}{Q(s)} = \frac{1}{As} \tag{4.8}$$

And Fig. 4.7(b) gives an electrical integral link with the transfer function:

$$G(s) = \frac{U_c(s)}{I(s)} = \frac{1}{cs} \tag{4.9}$$

According to the aforementioned two transfer functions, one can find that those two have the same form with the transfer function, although the two systems are different. For Fig. 4.7(c) and (d), the two systems are the hydraulic inertial link

Table 4.1: Analogue variables among different systems.

Mechanical translation system		**Electrical system**	
Variable	**Quantity**	**Variable**	**Quantity**
f	Force	U	Voltage
v	Velocity	I	Currant
M	Mass	L	Inductance
K	Stiffness coefficient	$1/C$	Reciprocal capacitance
B	Damping coefficient	R	Resistance

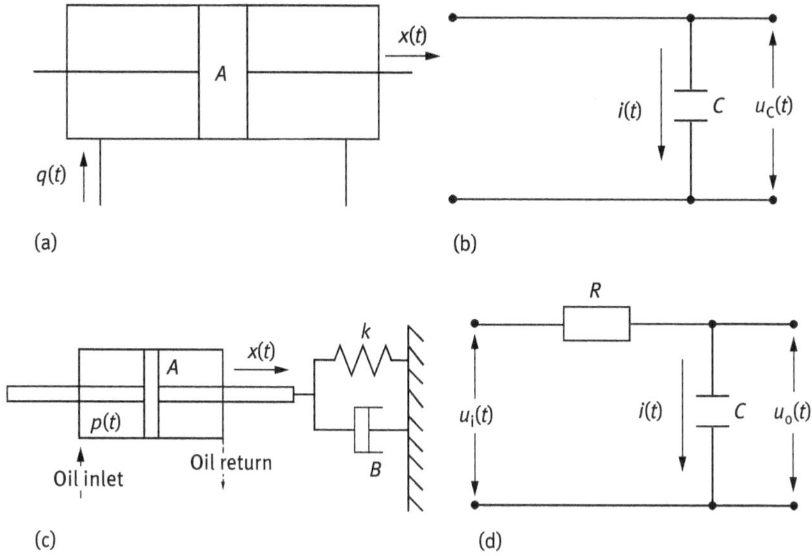

Fig. 4.7: Analogue relationship among different systems: (a) the hydraulic integral link; (b) the electrical integral link; (c) the hydraulic link: hydrocylinder-load system; and (d) the electrical inertial link.

and the electrical inertial link, respectively. The transfer functions of two different systems are eqs. (4.10) and (4.11), respectively. It is obvious that eqs. (4.10) and (4.11) have the same form except for the coefficients,

$$G(s) = \frac{X(s)}{P(s)} = \frac{A/k}{B/k \cdot s + 1} \tag{4.10}$$

$$G(s) = \frac{U_o(s)}{U_i(s)} = \frac{1}{RCs + 1} \tag{4.11}$$

Procedure to establish differential equation of a physical system is listed as follows:

Step 1: Analyze the principle of the practical system, make clear the relationships among all the parameters and the variables of the system, and define the system input and output variables.

Step 2: Start from the system input to express the dynamic equations of all the components in terms of the followed physical laws.

Step 3: Eliminate the intermediate variables so as to derive the differential equation that describes the relationship between the system input and output variable from the obtained linear dynamic equations.

Step 4: Normalize the differential equation. That is, to move all terms related to the input to the right-hand side of the differential equation and arrange in a descendent or an ascendant order, and to move all terms related to the output to the left-hand side of the differential equation and arrange in a descendent or an ascendant order.

4.4 Examples

Example 4.1

Apply Kirchhoff's law and Ohm's law to write a modeling equation for the system shown in Fig. 4.8. Here, $u_i(t)$ and $u_o(t)$ are the input voltage and output voltage, respectively. R_1, R_2, and C are the constants for every element of this system.

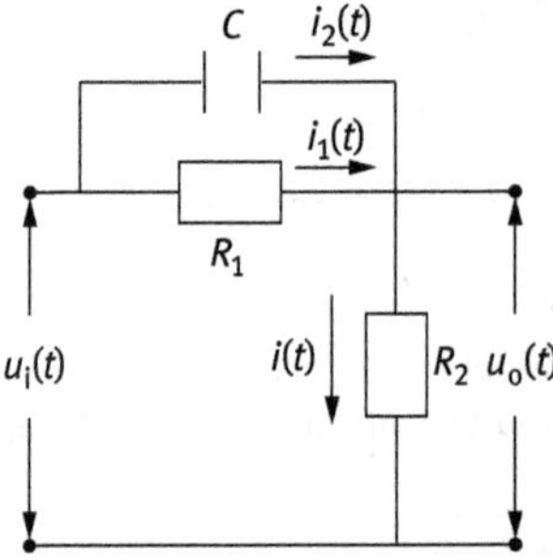

Fig. 4.8: A circuit system.

Solution

According to Kirchhoff's law and Ohm's law, together with nodal method of electrical network analysis, we can obtain

$$i_1(t) + i_2(t) = i(t)$$

$$u_i(t) = u_o(t) + R_1 i_1(t)$$

$$\frac{1}{C}\int i_2(t)\mathrm{d}t = R_1 i_1(t)$$

$$u_o(t) = R_2 i(t)$$

Eliminating the variables $i_1(t)$, $i_2(t)$, and $i(t)$ yields the differential equation for this circuit system:

$$R_1 C\frac{\mathrm{d}u_o(t)}{\mathrm{d}t} + \frac{R_1 + R_2}{R_2}u_o(t) = CR_1\frac{\mathrm{d}u_i(t)}{\mathrm{d}t} + u_i(t)$$

Example 4.2

Apply Kirchhoff's law and Ohm's law to write a modeling equation for the system shown in Fig. 4.9. Here, $u_i(t)$ and $u_o(t)$ are the input voltage and output voltage, respectively. R_1, R_2, L, and C are the constants for every element of this system.

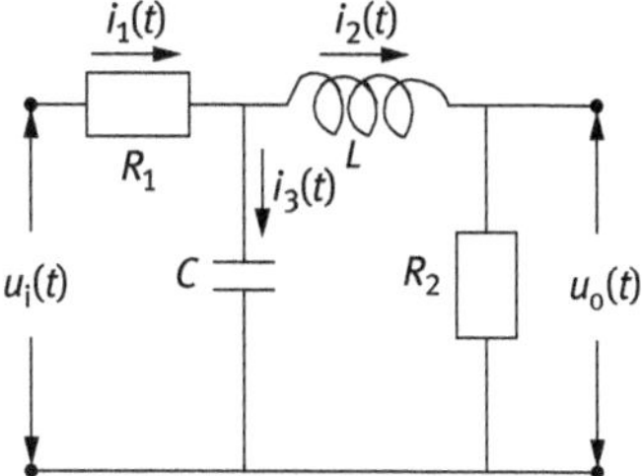

Fig. 4.9: A circuit system.

Solution

According to Kirchhoff's law and Ohm's law, together with nodal method of electrical network analysis, we can obtain

$$u_i = i_1 R_1 + \frac{1}{C}\int i_3 \mathrm{d}t$$

$$i_1 = i_2 + i_3$$

$$\frac{1}{C}\int i_3 \mathrm{d}t = L\frac{\mathrm{d}i_2}{\mathrm{d}t} + i_2 R_2$$

$$u_o = i_2 R_2$$

Eliminating the variables $i_1(t)$, $i_2(t)$, and $i_3(\mathrm{t})$ yields the differential equation,

$$R_1 LC\frac{\mathrm{d}^2 u_o(t)}{\mathrm{d}t^2} + (R_1 R_2 C + L)\frac{\mathrm{d}u_o(t)}{\mathrm{d}t} + (R_1 + R_2)u_o(t) = R_2 u_i(t)$$

4.5 Problems

P4.1. Write the equations describing the passive electrical circuit system shown in Fig. 4.10. The input variable is considered to be applied voltage $u_i(t)$, while the output variable is the voltage across the capacitor, $u_o(t)$.

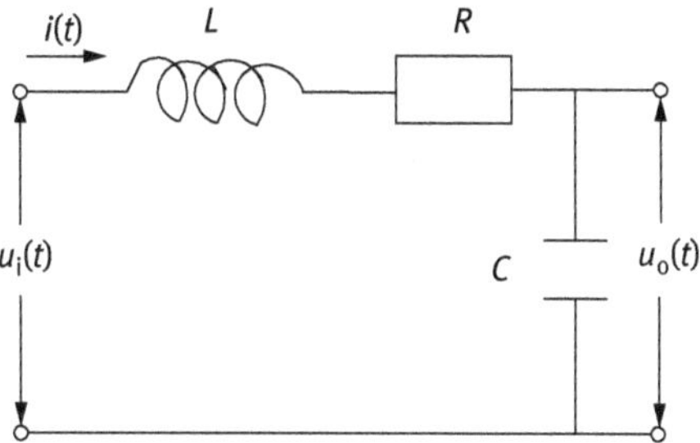

Fig. 4.10: *R–L–C* passive electrical circuit.

P4.2. Fig. 4.11 illustrates a filter, which is composed of two $R-C$ networks. Write the equations describing the passive electrical circuit.

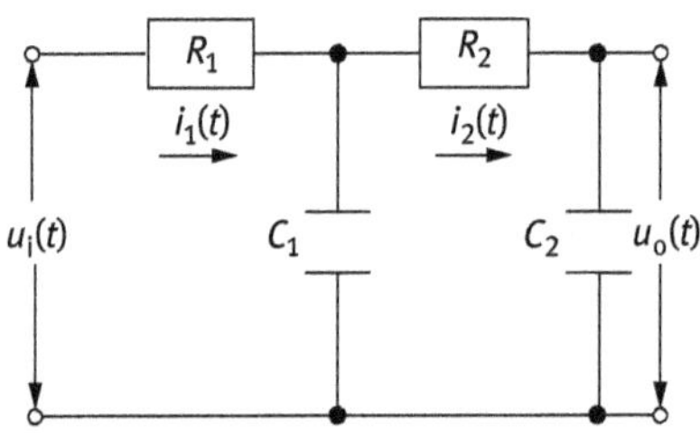

Fig. 4.11: Filter circuit.

P4.3. Apply Kirchhoff's law and Ohm's law to write the differential equation and transfer function for the system shown in Fig. 4.12. And u_i (t) and u_o (t) are the input voltage and output voltage, respectively. R_1, R_2, and C are the constants for every element of this system.

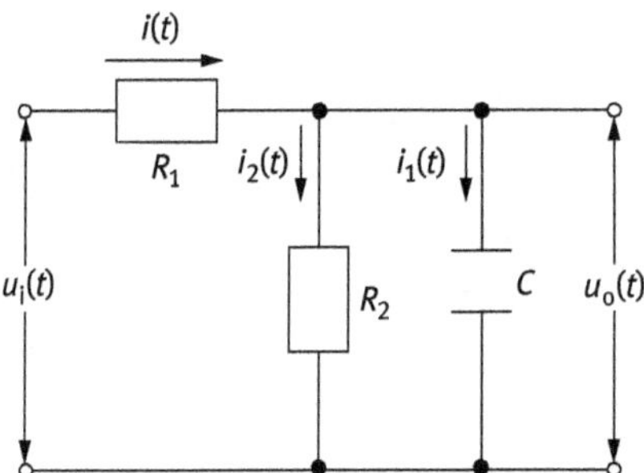

Fig. 4.12: The passive electrical circuit.

5 Fundamentals of Control Systems

5.1 Representation of Control Systems

The overall control system is commonly represented by the block diagram. Every block in the block diagram contains the transfer function (TF) of a component or element of the system.

There are two representations of control systems: the open-loop control system and closed-loop control system, as shown in Fig. 5.1(a), (b), respectively. We have described these two systems briefly in Chapter 1. Clearly, the open-loop control systems are special cases as that of closed-loop ones and are relatively simple to deal with. This and subsequent chapters are therefore concerned with closed-loop control systems, unless stated otherwise with some specific purposes.

5.2 The Transfer Function

In the analysis and design of linear systems, the TF and block diagrams comprise two of the most important tools. Block diagram reduction or simplification will be discussed at some length in Section 5.4. The reduction results in a single block containing a mathematical expression, which, for the moment, will be referred to as the TF of the whole system. Of course, for every part of the system, there are specific TFs depicting the typical links that will be described in detail in Section 5.3.

5.2.1 Definition of the TF

Let us consider, therefore, that the TF of a system or of an element is simply defined as the ratio of the Laplace transform of the output to the input, provided that all the initial conditions are zero. As it is a ratio of Laplace transforms, the TF is an algebraic function of s. Thus the TF should be

$$G(s) = \frac{F_o(s)}{F_i(s)} \tag{5.1}$$

where $F_o(s)$ and $F_i(s)$ are the Laplace transform of the output variable and the input variable, respectively.

A TF is nothing more than the s representation of a physical system that can be described by an ordinary differential equation with constant coefficients. In other words, consider now a system described by the general nth-order input–output equation:

https://doi.org/10.1515/9783110573275-005

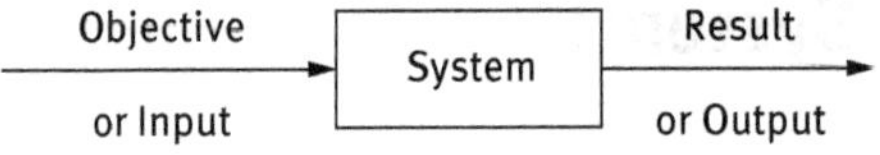

(a)

(b)

Fig. 5.1: Two representations of control systems: (a) the open-loop system and (b) the closed-loop system.

$$a_n y^{(n)} + a_{n-1} y^{(n-1)} + \cdots + a_1 \dot{y} + a_0 y = b_m x^{(m)} + b_{m-1} x^{(m-1)} + \cdots + b_1 \dot{x} + b_0 x \tag{5.2}$$

We assume that the input starts at $t = 0+$ and that $y(0), \dot{y}(0), \ldots, y^{(n-1)}(0)$, $x(0)$, $\dot{x}(0), \ldots, x^{(m-1)}(0)$ are all zero.

Then transform both sides of eq. (5.2), with the initial condition terms setting equal to zero. Collecting the remaining terms, we obtain the algebraic equation of s:

$$(a_n s^n + a_{n-1} s^{n-1} + \cdots + a_1 s + a_0) Y(s) = (b_m s^m + b_{m-1} s^{m-1} + \cdots + b_1 s + b_0) X(s)$$

which can be rearranged to give the transform of the output as

$$Y(s) = \left(\frac{b_m s^m + b_{m-1} s^{m-1} + \cdots + b_1 s + b_0}{a_n s^n + a_{n-1} s^{n-1} + \cdots + a_1 s + a_0} \right) X(s) \tag{5.3}$$

Hence the output transform $Y(s)$ is the product of the input transform $X(s)$ and a rational function of the complex variable s whose coefficients are the same as those in the input–output differential equation. This rational function of s is known as the system's TF. It plays a key role in the analysis of linear systems. According to eq. (5.3), denote the TF by $G(s)$:

$$G(s) = \frac{b_m s^m + b_{m-1} s^{m-1} + \cdots + b_1 s + b_0}{a_n s^n + a_{n-1} s^{n-1} + \cdots + a_1 s + a_0} \tag{5.4}$$

And we rewrite eq. (5.3) as

$$Y(s) = G(s) X(s) \tag{5.5}$$

Even if the system model is in a form other than the input–output differential equation, the TF can be found from eq. (5.5) or from its equivalent:

$$G(s) = \frac{Y(s)}{X(s)} \tag{5.6}$$

If the input is denoted by a general symbol instead of a specific function of time, then we can transform any version of the modeling equations, replace all of the initial condition terms by zero, and solve for the transformed output.

5.2.2 Properties of the TF

By concluding that similar systems have not only the same form of differential equations but also the same form of TFs, the following properties of TF are listed out:

(1) The TF is independent of the input to the system, i.e., the characteristics of the system are not modified by the input signal (at least within the working range).
(2) All initial conditions are assumed to be zero. The same thing is that the system is at rest initially.
(3) The TF can be applied to describe only time-invariant linear systems, i.e., systems whose parameters do not change or change only a little during operation.
(4) The unit of a TF is related to the units of the system input and output. A unit is not essential.

The first property is self-explanatory; the others, however, deserve a word of explanation.

The assumption of zero initial conditions is a property always associated with the use of Laplace transform. However, the Laplace transform of TF is still very useful, as many systems concerned us can be considered to be at rest initially.

Knowing that the Laplace transform is a kind of linear integral transform, the third property is undoubted. It is really a statement of two important generalizations, usually made of any "real" system that we may want to analyze. We assume that the system is both linear and time invariant, a situation which is rarely ever true in practice. For example, we use Ohm's law to describe how the current through a resistor varies in a constant linear manner directly with the voltage across the resistor. If, however, the voltage is too high, the resulting high current will cause heating in the resistor, thereby causing an increase in resistance, and eventually the resistor will melt. This "system," if operated out of its working range, becomes both time variant and nonlinear and eventually discontinuous. The resistor is designed to have a working range of voltage and current. While operated within its working range, its behavior can be assumed to be linear and time invariant. This argument applies to most systems that concern us, so we can use TFs to describe such system in most cases.

Example 5.1

Write TFs for the below differential equations (all initial conditions are zero):

$$\text{(1)}\quad 5\frac{d^3y}{dt^3}+2\frac{d^2y}{dt^2}+\frac{dy}{dt}+2y=6\frac{dx}{dt}+7x;$$
$$\text{(2)}\quad \frac{d^4y}{dt^4}+2\frac{d^3y}{dt^3}+6\frac{d^2y}{dt^2}+3\frac{dy}{dt}+2y=4x$$

Solution

Taking the Laplace transform according to the differential theorem under the initial conditions and then rearranging both sides of TF:

$$\text{(1)}\quad G(s)=\frac{Y(s)}{X(s)}=\frac{6s+7}{5s^3+2s^2+s+2};$$
$$\text{(2)}\quad G(s)=\frac{Y(s)}{X(s)}=\frac{4}{s^4+2s^3+6s^2+3s+2}$$

5.2.3 The Rational Polynomial Form of a TF

All the TFs derived so far have been in the form of constants and denominators as functions of s. It is possible, however, for TFs to have numerators as functions of s. A TF can thus be manipulated into the general form of a rational polynomial, such as in eq. (5.4):

$$G(s)=\frac{Y(s)}{X(s)}=\frac{b_m s^m+b_{m-1}s^{m-1}+\cdots+b_1 s+b_0}{a_n s^n+a_{n-1}s^{n-1}+\cdots+a_1 s+a_0}$$

where

a_n and b_m are real constants for real physical systems, $m<n$;
$Y(s)$ is the numerator polynomial; and
$X(s)$ is the characteristic polynomial of the system, which is also the denominator polynomial. And the equation $X(s)=0$ is the characteristic equation of the system.

At first glance, this statement seems to be odd with the simple systems we have just found. Actually it is not. TFs for the simple systems were derived using ideal mathematical models. These models are adequate enough for use in control applications where the frequencies generally are quite low. At extremely high frequencies, the models fail. These systems will eventually give no output as the frequency of the input becomes very high.

This expression can be factorized as

$$G(s) = K'\frac{(s - z_1)(s - z_2)\cdots(s - z_m)}{(s - p_1)(s - p_2)\cdots(s - p_n)} \tag{5.7}$$

where z_1, z_2,..., etc., are the roots of the equation $Y(s) = 0$, and p_1, p_2,..., etc., are the roots of the equation $X(s) = 0$. $K' = b_m/a_n$. The roots z_1, z_2, ..., etc., are called zeros; if $s = z_1, z_2$, ..., etc., the overall function will be zero. The roots p_1, p_2, ..., etc., are called poles; if $s = p_1, p_2$, ..., etc., the overall function will be infinite. $(s - z_i)$ is called a zero factor and $(s - p_i)$ is called a pole factor. z_i and p_i may be real or complex. The complex roots can now be expressed as complex numbers, e.g., $z_l = \sigma_l + j\omega_l$. In fact, the poles are the characteristic roots of the system. From the fact that a's and b's of the general equation must be real for any physical system and all complex roots must exist in conj ugate pairs, i.e., if $z_1 = -(\sigma + j\omega)$ and $z_2 = -(\sigma - j\omega)$, then

$$(s - z_1)(s - z_2) = (s + \sigma + j\omega)(s + \sigma - j\omega) = s^2 + 2\sigma s + (\sigma^2 + \omega^2) \tag{5.8}$$

while if $z_1 = -(\sigma_1 + j\omega_1)$ and $z_2 = -(\sigma_2 + j\omega_2)$, then

$$\begin{aligned}(s - z_1)(s - z_2) &= (s + \sigma_1 + j\omega_1)(s + \sigma_2 + j\omega_2)\\ &= s^2 + (\sigma_1 + j\omega_1 + \sigma_2 + j\omega_2)s + (\sigma_1 + j\omega_1)(\sigma_2 + j\omega_2)\end{aligned} \tag{5.9}$$

Equation (5.8) has real coefficients, while eq. (5.9) has possible imaginary coefficients and is therefore inadmissible. Since conjugate pairs of roots come from the solution of second-order equation, $G(s)$ can be rewritten as

$$G(s) = K'\frac{(s - z_1)(s - z_2)\cdots(s^2 + \beta_1 s + \alpha_1)(s^2 + \beta_2 s + \alpha_2)\cdots}{(s - p_1)(s - p_2)\cdots(s^2 + \delta_1 s + \gamma_1)(s^2 + \delta_2 s + \gamma_2)\cdots} \tag{5.10}$$

where all coefficients z, p, α, β, y, and δ are real.

The roots of $G(s)$ can be plotted on the "s-plane" diagram. For some system with the TF

$$G(s) = \frac{s + 2}{(s + 3)(s^2 + 2s + 2)}$$

We can plot the zeros and poles of this TF in the complex plane as shown in Fig. 5.2. Use the signal of "o" to represent zero, especially for $s_1 = -2$. And use the signal of "×" to represent the pole, especially for $s_2 = -3$ and $s_{3,4} = -1 \pm j$.

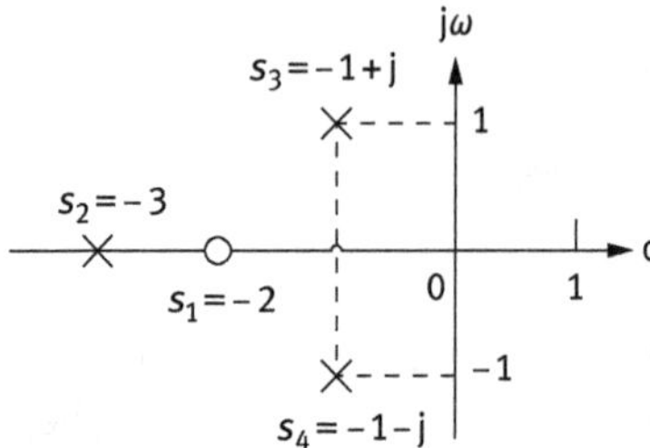

Fig. 5.2: Distribution diagram for zeros and poles.

5.2.4 TF of Elements in Series Connection

In order to analyze complicated feedback control systems, it is usual to rearrange the block diagram so as to enable it to be easily analyzed. In doing so, the objective is to produce a block diagram composed of some basic elements discussed below.

The TF of a system is generally defined as the ratio of the Laplace transformed output to the Laplace transformed input. Thus, it can be considered as a mathematical operator as shown in Fig. 5.3.

The elements are connected in series as shown in Fig. 5.4(a). The objective here is to find TF for the system in Fig. 5.4(a) such that it can be represented as that in Fig. 5.4(b). By applying the definition of TF for the system in Fig. 5.4(b), one has

$$G = \frac{Y}{X_1}$$

Fig. 5.3: Transfer function block of an element.

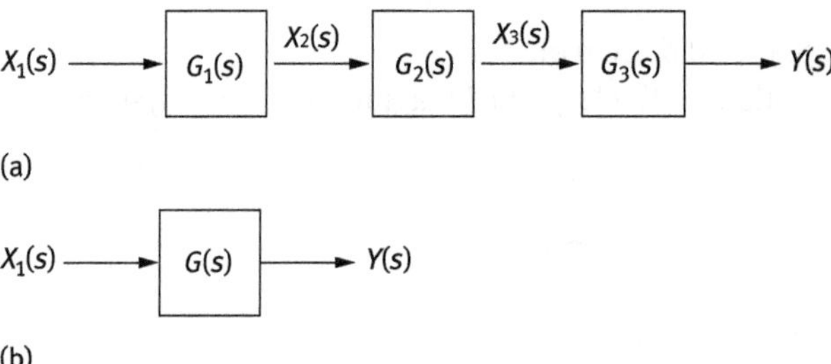

Fig. 5.4: Transfer function block of serial elements: (a) elements in series and (b) equivalent representation of (a).

which can be written as

$$G = \frac{Y}{X_1} = \frac{Y}{X_3}\frac{X_3}{X_2}\frac{X_2}{X_1} = G_3 G_2 G_1$$

Of course, one can generalize the above expression for n elements connected in series so that TF becomes

$$G = G_1 G_2 G_3 \cdots = \prod_{i=1}^{n} G_i \tag{5.11}$$

5.2.5 TF of Elements in Parallel Connection

In this case, the elements are connected in parallel as shown in Fig. 5.5(a). The objective is to find the TF for the system in Fig. 5.5(a) such that it can be represented as that in Fig. 5.5(b).

To find the TF for this system as shown in Fig. 5.5(b), one can apply the definition again so that

$$G = \frac{Y}{X}$$

But the output signal is

$$Y = Y_1 + Y_2 + Y_3 = G_1 X + G_2 X + G_3 X = (G_1 + G_2 + G_3)X$$

Therefore, upon generalization of the above expression for n elements connected in parallel one has the TF of the system as

$$G = \sum_{i=1}^{n} G_i \tag{5.12}$$

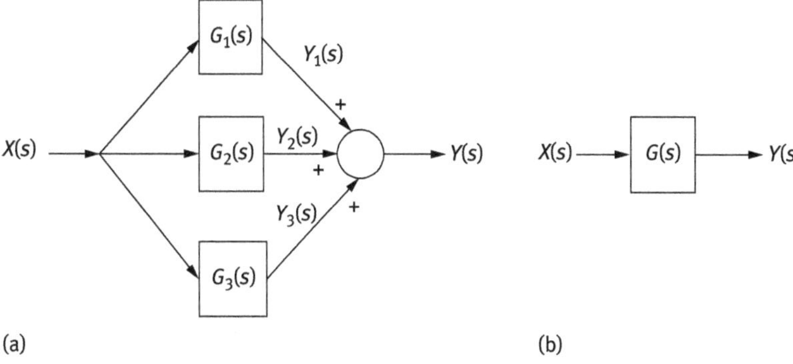

Fig. 5.5: Equivalent representation for parallel connection: (a) elements in parallel and (b) equivalent representation of (a).

5.2.6 Remarks

(1) A general engineering control system usually contains a combination of elements in series as well as elements in parallel connections.

(2) Two basic assumptions for the TF representation and solution are passivity (i.e., the individual element has no energy source) and linearity (such that the principle of superposition can be applied).

Example 5.2

Evaluate the TFs $Y(s)/U(s)$ and $Z(s)/U(s)$ for the block diagram shown in Fig. 5.6, giving the results as rational functions of s.

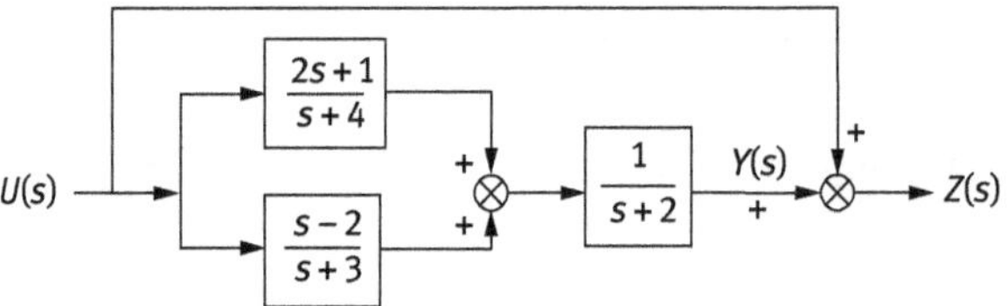

Fig. 5.6: Blocks diagram for Example 5.2.

Solution

Because $Z(s)$ can be viewed as the sum of the outputs of two parallel blocks, one of which takes $Y(s)$ as its output, so we first evaluate the TF $Y(s)/U(s)$. To do this, we observe that $Y(s)$ can be considered as the output of a series combination of two parts, and one of which is a parallel combination of two blocks. Starting with this parallel combination, we write

$$\frac{2s+1}{s+4}+\frac{s-2}{s+3}=\frac{3s^2+9s-5}{s^2+7s+12}$$

and redraw the block diagram as shown in Fig. 5.7(a). The series combination in this version has the TF

$$\frac{Y(s)}{U(s)}=\frac{3s^2+9s-5}{s^2+7s+12}\cdot\frac{1}{s+2}=\frac{3s^2+9s-5}{s^3+9s^2+26s+24}$$

which leads to the diagram shown in Fig. 5.7(b). We can reduce the final parallel combination to the single block shown in Fig. 5.7(c) by writing

$$\frac{Z(s)}{U(s)}=1+\frac{Y(s)}{U(s)}=1+\frac{3s^2+9s-5}{s^3+9s^2+26s+24}=\frac{s^3+12s^2+35s+19}{s^3+9s^2+26s+24}$$

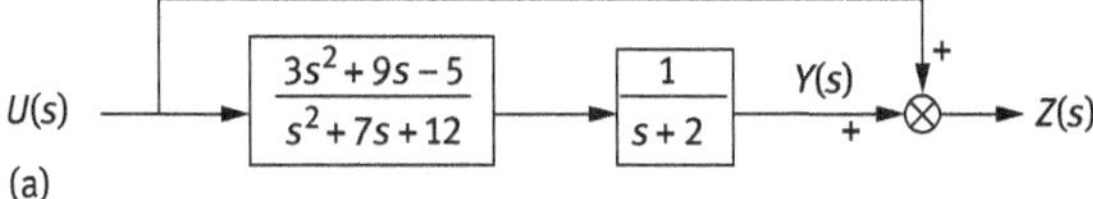

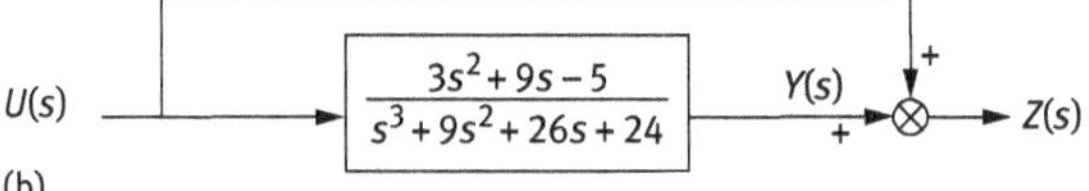

U(s)
(a)
3s²+9s−5 / s²+7s+12
1 / s+2
Y(s)
+
+
Z(s)
U(s)
(b)
3s²+9s−5 / s³+9s²+26s+24
Y(s)
+
+
Z(s)
U(s)
(c)
s³+12s²+35s+19 / s³+9s²+26s+24
Z(s)

Fig. 5.7: Equivalent block diagrams for the diagram: (a) Combine two parallel links. (b) Combine two series links. (c) Combine the other two parallel links.

5.3 The TF for Typical Links

In mechanical engineering, numerous normal systems consist of some typical TFs, although their physical structures and working principles are different. We call those typical TFs as links. Link does not represent a kind of component. Sometimes link is made up of several components, which is called a part, unit, or system. Link even means a kind of mechanism sometimes. Perfectly grasping every typical link will be very helpful for analyzing and researching complicated systems.

The general form of the TF is

$$G(s)=\frac{K\prod_{i=1}^{b}(\tau_i s+1)\prod_{l=1}^{c}\left(\tau_l^2 s^2+2\zeta_l\tau_l s+1\right)}{s^{\nu}\prod_{j=1}^{d}(T_j s+1)\prod_{k=1}^{e}\left(T_k^2 s^2+2\zeta_k T_k s+1\right)} \tag{5.13}$$

where

$$K=\frac{b_m}{a_n}\cdot\prod_{i=1}^{b}\frac{1}{\tau_i}\cdot\prod_{l=1}^{c}\frac{1}{\tau_l^2}\cdot\prod_{j=1}^{d}T_j\cdot\prod_{k=1}^{e}T_k^2$$

In eq. (5.13) one can find every link for system directly and we will explain in detail about these links, noting that the connection of every link is in series.

Referring to Fig. 4.7, every link is named according to its differential equation rather than its physical device or component's name. The link is based on the characteristic of the motion caused by several components. The same component has a different role in different system. Even in the same system, different input and output variables can lead the same component function directly, just like different link.

5.3.1 Proportional Link

The differential equation for proportion link is

$$y(t) = kx(t)$$

After taking Laplace transform of both sides for the above equation and rearranging, the TF of proportion link is

$$G(s) = \frac{Y(s)}{X(s)} = k \tag{5.14}$$

where k is the proportional constant. The proportion link exists in many fields, such as the output speed and the input speed in a gear system, the output displacement and the input displacement in a lever system, the output voltage and the input rotated angle of a potentiometer, and the output signal and the input signal of an electronic amplifier.

5.3.2 Integral Link

The differential equation for integral link is

$$y(t) = k\int x(t)\mathrm{d}t$$

So the TF of integral link is

$$G(s) = \frac{Y(s)}{X(s)} = \frac{k}{s} \tag{5.15}$$

where k is the integral constant.

Example 5.3a
The schematic diagram of the hydrocylinder is shown in Fig. 5.8. The cross-sectional area is A, the input flow is $q(t)$, and the output is the velocity $v(t)$ of the hydrocylinder's piston. Find the TF for this system. (All initial conditions are zero).

Solution
The velocity of the piston is

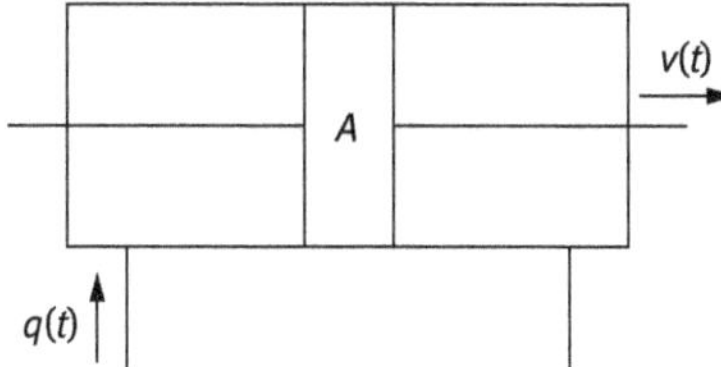

Fig. 5.8: The hydraulic proportion link.

$$v(t) = \frac{q(t)}{A}$$

Taking Laplace transform of the above equation and rearranging it to get the TF, we have

$$G(s) = \frac{V(s)}{Q(s)} = \frac{1}{A}$$

Obviously, it's a hydraulic proportion link.

Example 5.3b
The schematic diagram of the hydrocylinder is shown in Fig. 5.9. The cross-sectional area is A, the input flow is $q(t)$, and the output is the displacement $x(t)$ of the hydrocylinder's piston. Find the TF for this system.

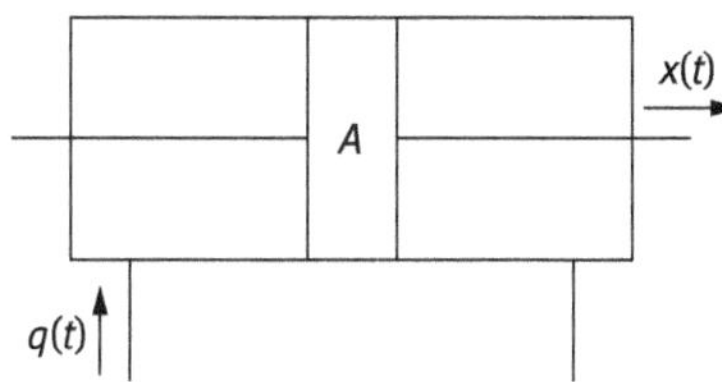

Fig. 5.9: The hydraulic integral link.

Solution
The velocity of the piston is

$$v(t) = \frac{q(t)}{A} = \frac{\mathrm{d}x(t)}{\mathrm{d}t}$$

Taking Laplace transfer of the above equation and rearranging it to get the TF, we have

$$G(s) = \frac{X(s)}{Q(s)} = \frac{1}{As}$$

Compared with eq. (5.15), it's the integral link.

Summary from above examples

(1) We conclude from the above examples that the link is integral when output is displacement and input is flow, but it is a proportion when output changes into velocity and input is the same thing.

(2) For a physical system, the type of the link is based on the TF depicted by the input and the output. In other words, what the input and output are or which one is the input and output has an important effect.

Example 5.4

The schematic diagram of the passive circuit is shown in Fig. 5.10. The input is the current $i(t)$, and the output is the voltage $u_c(t)$. Find the TF for this passive circuit.

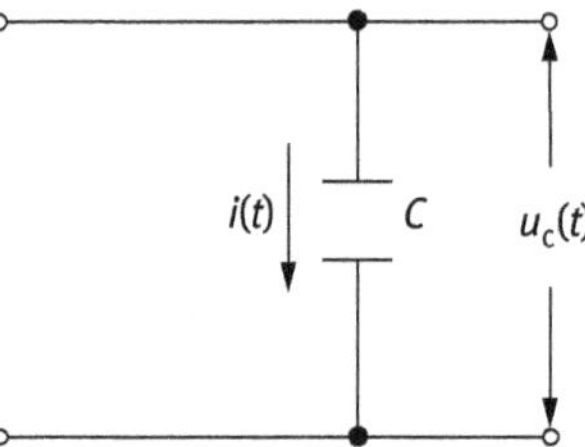

Fig. 5.10: The electrical integral link.

Solution

The differential equation for the current $i(t)$ through the capacitor and the voltage across the capacitor is

$$u_c(t) = \frac{1}{C}\int i(t)\mathrm{d}t$$

Taking Laplace transform of the above equation and rearranging it to get the TF, we have

$$G(s) = \frac{U_c(s)}{I(s)} = \frac{1}{Cs}$$

Obviously, it's an electrical integral link.

5.3.3 Inertial Link

The differential equation for inertial link is

$$T\frac{\mathrm{d}y(t)}{\mathrm{d}t} + y(t) = Kx(t)$$

Taking Laplace transform of the above equation, we have

$$TsY(s) + Y(s) = KX(s)$$

After rearranging the above equation, the TF of inertial link is

$$G(s) = \frac{Y(s)}{X(s)} = \frac{K}{Ts+1} \tag{5.16}$$

where T is the time constant which may be a constant for a spring and K is a coefficient of some element which may be an area for a piston.

Example 5.5

The schematic diagram of the passive circuit is shown in Fig. 5.11. The input voltage is $u_i(t)$ and the output voltage is $u_o(t)$. C and R are constants of the capacitor and resistor, respectively. $i(t)$ is just an intermediate variable and should not exist in the final TF. Find the transfer function for this passive circuit.

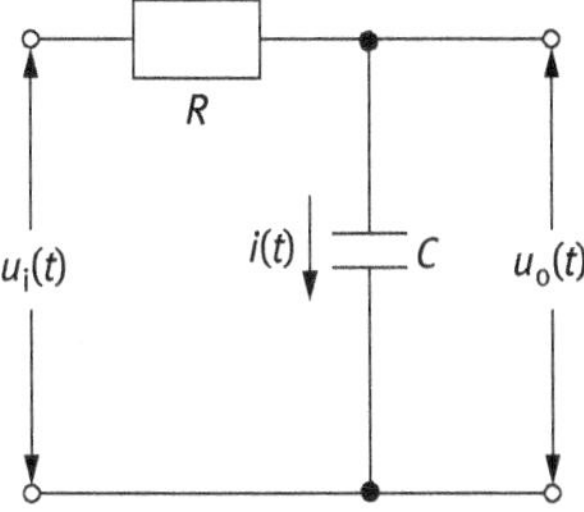

Fig. 5.11: The electrical inertial link.

Solution

According to Kirchhoff's law, to build the differential equations for this passive circuit

$$\begin{cases} u_i(t) = i(t)R + u_o(t) \\ u_o(t) = \frac{1}{C}\int i(t)\mathrm{d}t \end{cases}$$

Cancel the current variable $i(t)$, and rearrange,

$$u_i(t) = RC\frac{\mathrm{d}u_o(t)}{\mathrm{d}t} + u_o(t)$$

Taking Laplace transform of the above equation and then rearranging it, we have

$$G(s) = \frac{U_o(s)}{U_i(s)} = \frac{1}{RCs+1} = \frac{1}{Ts+1}$$

where T is the time constant with the value of RC. According to eq. (5.16), it's an electrical inertial link.

Example 5.6

The hydraulic system is shown in Fig. 5.12, which control the spring and the damping. K and B are constants for spring and damping, respectively. The input $p(t)$ is the inlet pressure and the output $x(t)$ is the displacement of the piston. A is the working area of the piston. Find the TF for this hydrocylinder-load system.

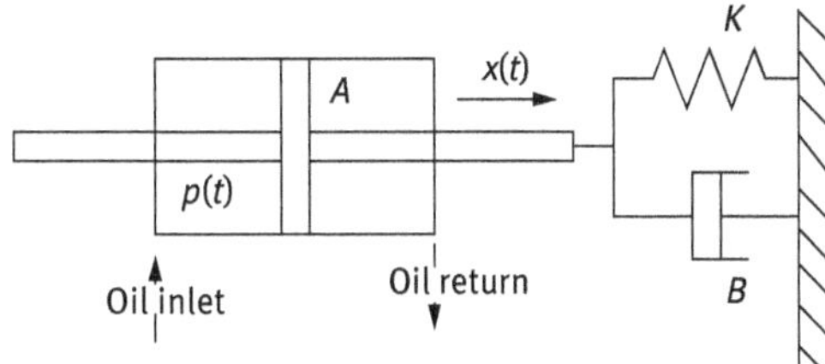

Fig. 5.12: The hydraulic inertial link.

Solution

First, according to the principle of the hydrocylinder, we have

$$F(t) = Ap(t)$$

At the same time, the hydraulic pressure force $F(t)$ balances the load of the damping and the spring,

$$F(t) = B\frac{dx(t)}{dt} + Kx(t)$$

Combining the above two equations to get the differential equation for the hydrocylinder

$$B\frac{dx(t)}{dt} + Kx(t) = Ap(t)$$

Taking Laplace transform of the above equation and then rearranging it, we have

$$G(s) = \frac{X(s)}{P(s)} = \frac{A}{Bs + K} = \frac{A/K}{B/K \cdot s + 1}$$

Obviously, it's a hydrocylinder inertial link.

5.3.4 Differential Link

There are three kinds of differential links: ideal differential link, first-order differential link, and second-order differential link. The equations of those three kinds of differential links are as follows:

$$y(t) = T\frac{dx(t)}{dt}$$

$$y(t) = T\frac{dx(t)}{dt} + x(t)$$

$$y(t) = T^2\frac{d^2x(t)}{dt^2} + 2\zeta T\frac{dx(t)}{dt} + x(t)$$

At the same time, the TFs for the above three differential equations are

$$G(s) = \frac{Y(s)}{X(s)} = Ts \tag{5.17}$$

$$G(s) = \frac{Y(s)}{X(s)} = Ts + 1 \tag{5.18}$$

$$G(s) = \frac{Y(s)}{X(s)} = T^2s^2 + 2\zeta Ts + 1 \tag{5.19}$$

where T is the time constant, and ζ is the damping ratio.

If $T^2s^2 + 2\zeta Ts + 1 = 0$ has two real roots, the equation of $T^2s^2 + 2\zeta Ts + 1 = 0$ is not the second-order differential link. In fact, it is a system that has two first-order differential links connected in series.

Example 5.7

The electrical link is shown in Fig. 5.13. The input voltage is $u_i(t)$ and the output voltage is $u_o(t)$. C and R are constants of the capacitor and resistor, respectively. $i(t)$ is just an intermediate variable and should not exist in the final TF. Find the TF for this electrical link.

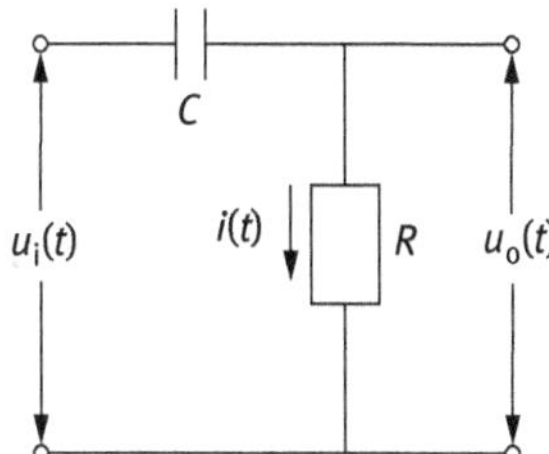

Fig. 5.13: The electrical differential link.

Solution

According to Kirchhoff's law and Ohm's law to build the differential equation

$$\begin{cases} u_i(t) = \frac{1}{C}\int i(t)\mathrm{d}t + u_o(t) \\ i(t) = u_o(t)/R \end{cases}$$

Eliminate the intermediate variable $i(t)$:

$$u_i(t) = \frac{1}{C}\int \frac{u_o(t)}{R}\mathrm{d}t + u_o(t)$$

Finally, get the TF for this electrical differential link:

$$G(s) = \frac{U_o(s)}{U_i(s)} = \frac{RCs}{RCs + 1} = \frac{Ts}{Ts + 1}$$

If the value of $T = RC$ is very small, the above equation is rewritten approximately as $G(s) = Ts$ as Ts in the denominator can be neglected. On this occasion, the electrical system could be regarded as an ideal differential link.

5.3.5 Oscillation Link

The differential equation for oscillation link is

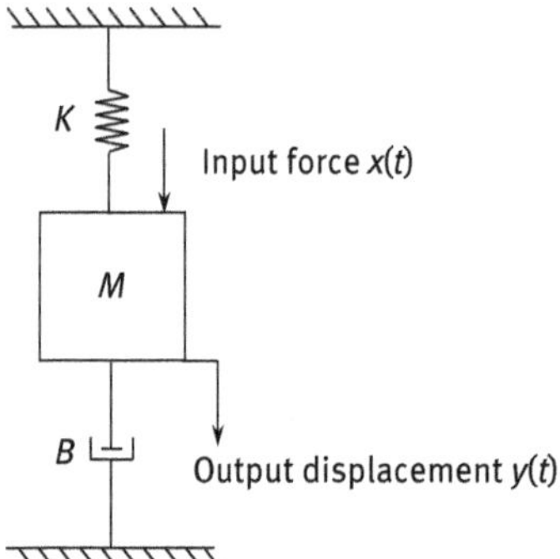

Fig. 5.14: *K–M–B* second-order oscillation link.

$$T^2\frac{\mathrm{d}^2y(t)}{\mathrm{d}t^2}+2\zeta T\frac{\mathrm{d}y(t)}{\mathrm{d}t}+y(t)=Kx(t)$$

Taking Laplace transform of the above equation to get the TF, we have

$$G(s)=\frac{Y(s)}{X(s)}=\frac{K}{T^2s^2+2\zeta Ts+1} \tag{5.20}$$

Taking the translational mechanical system as an example, as shown in Fig. 5.14, the differential equation is

$$M\frac{\mathrm{d}^2y(t)}{\mathrm{d}t^2}+B\frac{\mathrm{d}y(t)}{\mathrm{d}t}+Ky(t)=x(t)$$

Then, take the Laplace transform of the above differential equation and get the TF:

$$G(s)=\frac{Y(s)}{X(s)}=\frac{1}{Ms^2+Bs+K}$$

Obviously, this system is the second-order oscillation link.

5.4 Block Diagrams

5.4.1 Introduction

Any engineering system can be represented by a combination of blocks. Each block has a single line inside and a single line outside. Within the block is a statement of its operation, that is, the block indicates what happens to the input information after transmission. Figure 5.15 shows this process. In the illustration, a simple amplifier is shown as a block. If, for example, the amplifier has a gain of 200 and a 1 mV signal is fed into it, then the amplifier will produce 200 mV at its output terminal.

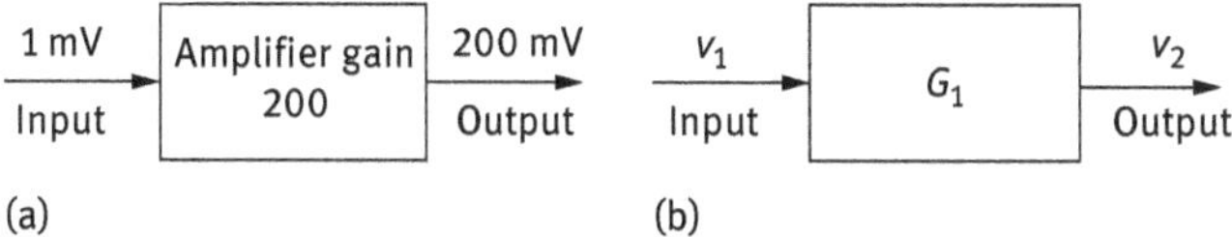

Fig. 5.15: A block diagram of an amplifier: (a) precise representation and (b) general representation.

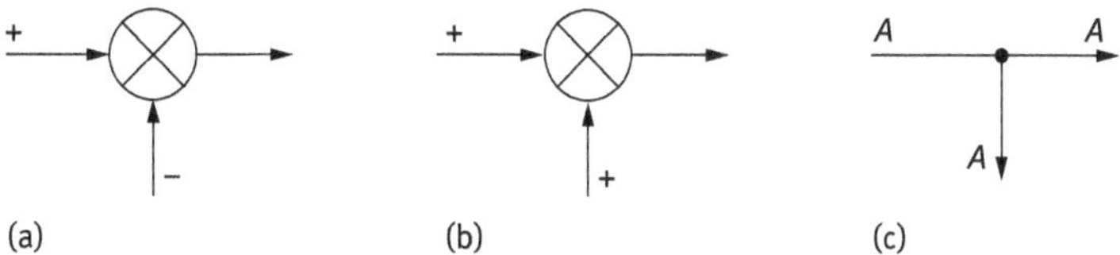

Fig. 5.16: Summing point and tie point: (a) negative summing point; (b) positive summing point; and (c) tie point.

5.4.2 Summing Point and Tie Point

It is sometimes necessary to feed more than one signal into an amplifier at the same time and this is done by a summing network in an actual circuit. However, in block diagram it is done by means of a summing junction or summing point, as shown in Fig. 5.16(a), (b). A summing junction is always shown as a circle with arrows into and out of the symbol. The arrows into the symbol are always identified with a plus or minus sign, indicating either a positive or a negative signal. The signal out of the summing point means algebraic sum of the signals into it.

The manipulation (or equivalent transform) of summing point and block as shown in Table 5.1 (rules 4 and 5) will be proved in later sections.

In block diagrams, it is often necessary to have the signal flow in more than one direction. To affect this symbolically, a tie point or pick off point is used in Fig. 5.16(c). A tie point can be thought of as a node or junction where no summing (positively or negatively) is done. It is just a convenient method of providing extra paths along with the signal that may flow without affecting its original path. Tie point, like summing junction, may be transferred across blocks, and the same rules can be used as the movement of summing junction. These rules are illustrated in Table 5.1 (rules 6 and 7).

5.4.3 Terminologies

In order to make it easier to learn the following sections, we should first understand some useful and important terminologies. Referring to Fig. 5.17.

(1) Forward path is the path where the signal passes nonrepeatedly from input to output.

Table 5.1: Connections and equivalent transforms of block diagrams.

Rule	Connections	Original block diagrams	Equivalent block diagrams
1	Combining serial blocks	R, G_1, G_2, C	R, G_1G_2, C
2	Combining parallel blocks	R, G_1, G_2, +, ±, C	R, $G_1 \pm G_2$, C
3	Closing a feedback loop	R, +, ±, G, H, C	R, $\frac{G}{1 \mp GH}$, C
4	Moving a summing junction to the front of a block	R, G, +, ±, C, B	R, +, ±, G, $1/G$, C, B
5	Moving a summing junction to the back of a block	R, +, ±, B, G, C	R, G, +, ±, C, B, G
6	Moving a tie point to the front of a block	R, G, C, B	R, G, C, G, B
7	Moving a tie point to the back of a block	R, G, C, B	R, G, C, $1/G$, B

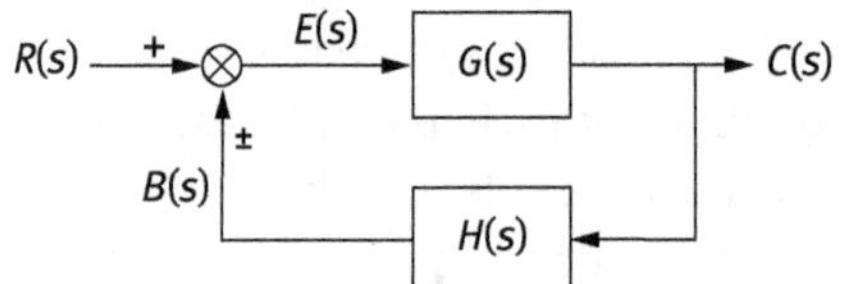

Fig. 5.17: Basic block diagram of a closed-loop system.

(2) In the forward path, the product of every link passed with the expression of $G(s)$ is the TF:

$$G(s) = \prod_{i=1}^{n} G_i(s) \tag{5.21}$$

(3) Feedback path is the path where the signal starts from the output and ends with the input.

(4) In the feedback path, the product of every link passed with the expression of $H(s)$ is the TF:

$$H(s) = \prod_{j=1}^{m} H_j(s) \tag{5.22}$$

(5) Open-loop TF of the system is the ratio of the primary feedback signal $B(s)$ to error signal $E(s)$:

$$G_{\text{open}}(s) = \frac{B(s)}{E(s)}$$

According to Fig. 5.17, we have

$$\begin{aligned}\frac{Y(s)}{X(s)} &= \frac{E(s) \cdot G(s)}{E(s) \pm B(s)} = \frac{G(s)}{1 \pm (B(s)/E(s))} = \frac{G(s)}{1 \pm ((Y(s)H(s))/E(s))} \\ &= \frac{G(s)}{1 \pm ((E(s) \cdot G(s)H(s))/E(s))} = \frac{G(s)}{1 \pm G(s)H(s)}\end{aligned}$$

So in terms of the above deduction, the open-loop TF is

$$G_{\text{open}}(s) = \frac{B(s)}{E(s)} = G(s)H(s) = \prod_{i=1}^{n} G_i(s) \prod_{j=1}^{m} H_j(s) \tag{5.23}$$

(6) Substituting eqs. (5.21) and (5.22) into the expression of $Y(s)/X(s)$, we have the closed-loop TF of the system:

$$\frac{Y(s)}{X(s)} = \frac{\prod_{i=1}^{n} G_i(s)}{1 \pm \prod_{i=1}^{n} G_i(s) \prod_{j=1}^{n} H_j(s)} \tag{5.24}$$

And when $H(s) = 1$, we call it a unity feedback system:

$$\frac{Y(s)}{X(s)} = \frac{\prod_{i=1}^{n} G_i(s)}{1 \pm \prod_{i=1}^{n} G_i(s)}$$

Taking Fig. 5.17 as an example again, we could find the relationship between the closed-loop TF and the open-loop TF. As illustrated in Fig. 5.17, the closed-loop TF is

$$\Phi(s) = \frac{G(s)}{1 \pm G(s)H(s)}$$

In order to simplify, we just select a unit negative feedback system to analyze the relationship. So, the above equation is simplified into

$$\Phi(s) = \frac{G(s)}{1 + G(s)}$$

where $G(s)$ is the open-loop TF in denominator and is also the forward path TF in numerator in fact.

After rearranging the equation about the closed-loop TF $\Phi(s)$, we could get the equation about the open-loop TF:

$$G(s) = \frac{\Phi(s)}{1 - \Phi(s)}$$

With the help of the relationship between the closed-loop TF and the open-loop TF, one could get the open-loop TF easily according to the closed-loop TF, especially for some calculation in the following chapters.

5.4.4 Simplification of the Block Diagram

There are several steps to simplify the procedure obtaining the closed-loop TF of a complicated control system. In the following sections, some commonly used techniques are introduced and some examples about block diagram reduction are also subsequently included.

5.4.4.1 Moving Tie Points

See Table 5.1 (rules 6 and 7): this method moves the tie point of a signal to front or back of G, which is illustrated in Fig. 5.18.

With reference to the block diagram on the left-hand side (LHS) of Fig. 5.18(a), the tie point is in front of G. Then

$$R = B$$

If the tie point is in back of G as indicated on the RHS block diagram of Fig. 5.18(a), then from the signal R to B must exist a TF $1/G$:

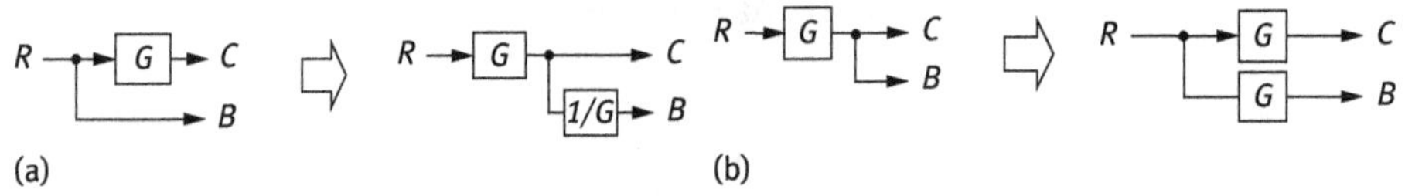

R G C B ⇨ R G C 1/G B
(a)

R G C B ⇨ R G C G B
(b)

Fig. 5.18: Equivalent block diagrams before and after moving a tie point: (a) moving a tie point to the back of a block and (b) moving a tie point to the front of a block.

$$R \cdot G \cdot \frac{1}{G} = B \Rightarrow R = B$$

We get the same signal B in both block diagrams of Fig. 5.18. In other words, the block diagram on the RHS of Fig. 5.18(a) is equivalent to the block diagram on the LHS of Fig. 5.18(a). The situation in Fig. 5.18(b) is the same about principle with Fig. 5.18(a), just different about moving order.

Example 5.8

Figure 5.19(a) shows a system with two feedback loops. To obtain the block diagram with a single feedback loop, use Fig. 5.18 and the rules for combining blocks in series and parallel.

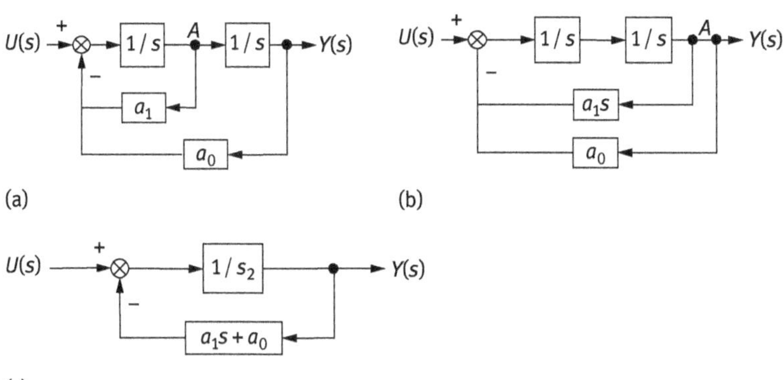

Fig. 5.19: Simplifying processes for Example 5.8: (a) block diagram for Example 5.8; (b) moving tie point A; and (c) find the internal feedback transfer function.

Solution

The feedback path from the output variable to the input variable is always named primary feedback path. Besides the primary feedback path, the other feedback path is named as inner feedback path. The tie point A for the inner feedback path can be moved to the output $Y(s)$ by replacing a_1 by a_1s, using rule 7 in Table 5.1, as shown in Fig. 5.19(b). Then the two integrator blocks, which are now in series, can be combined to give the TF $1/s^2$, using rule 1 in Table 5.1. The two feedback blocks are now in parallel and can be combined into the single TF $a_1s + a_o$, rule 2 in Table 5.1. With these changes, the revised diagram has a single feedback loop as shown in Fig. 5.19(c).

5.4.4.2 Moving Summing Points

This method is illustrated in Fig. 5.20(a) and (b) (see Table 5.1 (rules 4 and 5)).

With reference to the block diagram on the LHS of Fig. 5.20(a), one has

$$C = (R \pm B)G = RG \pm BG$$

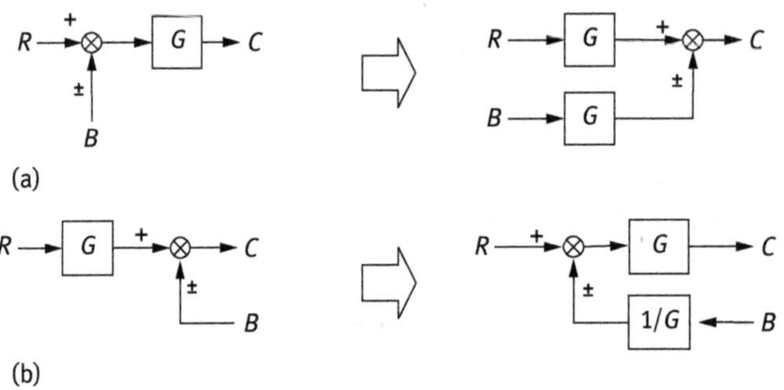

Fig. 5.20: Equivalent block diagrams before and after moving a summing point: (a) moving a summing junction to the back of a block and (b) moving a summing junction to the front of a block.

This is equivalent to the block diagram on the RHS of Fig. 5.20(a) since

$$C = RG \pm BG$$

The situation in Fig. 5.18(b) is the same of the principle in Fig. 5.18(a), but different in the moving order.

Example 5.9

Use Fig. 5.20 and the rule for a parallel combination to remodel the block diagram in Fig. 5.21(a) by removing right summing junction and leaving only left summing junction.

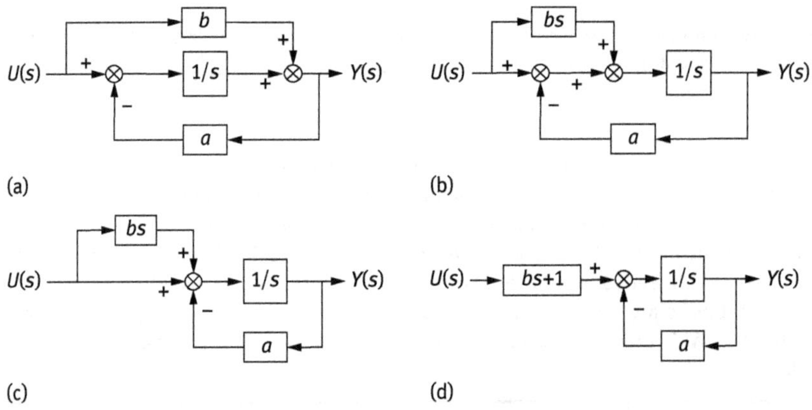

Fig. 5.21: Simplifying processes of Example 5.9: (a) block diagram for Example 5.9; moving the summing point; (c) combining the summing point; and (d) combining the two parallel paths.

Solution

We use Fig. 5.20(b) to move the right summer behind the integrator block to the front of integrator, resulting in Fig. 5.21(b). Since no block lies between the two summing junctions, the two summing junctions can be combined as a single summer, as shown in Fig. 5.21(c). The final step is to combine the two parallel paths in front of $U(s)$, resulting in Fig. 5.21(d), where the TF $(bs + 1)$ appears explicitly.

The diagram in Fig. 5.21(d) may not appear to be much simpler than the original diagram. However, it has the desirable property that the summing junction that entered the feedback loop after the integrator has been removed from the loop. Being discussed in the following section, it is a simple task to simplify the remaining feedback loop into a single block whose TF can be written as a ratio of polynomials. Once that has been done, the rule for combining blocks in series can be used to obtain a single overall TF from the input $U(s)$ to the output $Y(s)$:

$$\frac{Y(s)}{U(s)} = (bs+1)\cdot\frac{1/s}{1+(a/s)} = \frac{bs+1}{s+a}$$

5.4.4.3 General Feedback Case

However, the error signal $E(s)$ and the feedback signal $B(s)$ are given by (see Fig. 5.17)

$$\begin{cases} E(s) = R(s) \mp B(s) \\ B(s) = H(s)\cdot C(s) \end{cases}$$

So

$$E(s) = R(s) \mp B(s) = R(s) \mp H(s)\cdot C(s) \tag{5.25}$$

Also we have

$$C(s) = G(s)\cdot E(s) \tag{5.26}$$

Substitute eq. (5.25) into eq. (5.26)

$$C(s) = G(s)\cdot E(s) = G(s)\cdot[R(s) \mp H(s)C(s)]$$

Simplify the above equation and get the final TF for the classic closed-loop system:

$$\frac{C(s)}{R(s)} = \frac{G(s)}{1 \pm G(s)H(s)}$$

Also

$$\frac{E(s)}{R(s)} = \frac{1}{1 + G(s)H(s)}$$

The open-loop TF, independent of where the loop is opened, is $G(s)H(s)$. Obviously for a given open-loop TF, there can only be one resulting closed-loop TF as given by eq. (5.24). This fact is of major significance since it means that a complete knowledge of the open-loop system gives a complete knowledge of the closed loop. The problem of system stability will be dealt with later, but it is clear that if a closed-loop system is unstable, no measurements can possibly be made upon it; while open-loop measurements are quite practical, they enable a designer to see what compensation is necessary to make the closed-loop stable.

In order to move individual blocks across summing junctions or tie points, there are a number of other useful rules, and these are summarized in Table 5.1. Most of them are self-explanatory, and their use will be illustrated by examples. In general, the simplification of complex block diagrams may be accomplished by systematic modification of the diagram using the rules in Table 5.1 in the following order:

Step 1: Combine all serial blocks (rule 1).

Step 2: Combine all parallel blocks (rule 2).

Step 3: Close all inner loops (rule 3).

Step 4: Move summing junctions to the left or right of a block and tie points to the left or right (rules 4–7).

Example 5.10

Use the rules in Table 5.1 to simplify the block diagram in Fig. 5.22(a) and get the final TF of this system.

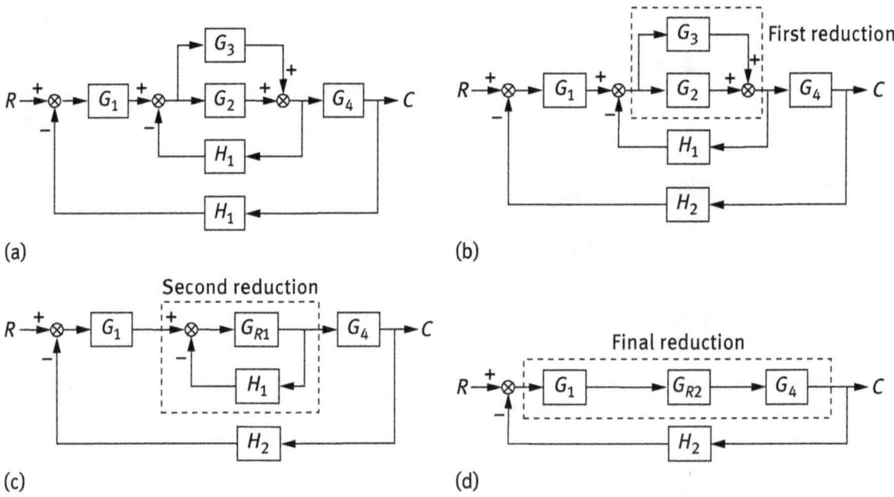

Fig. 5.22: Simplifying processes for Example 5.10: (a) the block diagram for Example 5.10; (b) control system with first block diagram reduction; (c) second block diagram reduction; and (d) final feedback control system.

Solution

In this example, the closed-loop TF of the control system is determined by applying the block diagram reduction rules in Table 5.1.

1) First reduction
With reference to the control system shown in Fig. 5.22(b), the two TFs in parallel connection and boxed with dotted line are first considered. By eq. (5.12), the TF of the first reduction is

$$G_{R1} = G_2 + G_3$$

2) Second reduction
Applying eq. (5.24) to the inner closed loop in Fig. 5.22 (c) so that the TF is

$$G_{R2} = \frac{G_{R1}}{1 + G_{R1}H_1}$$

Then, the forward loop TF in Fig. 5.22(c) becomes

$$G = G_1 G_{R1} G_4$$

3) Final block diagram reduction
Applying eq. (5.24) to the feedback loop in Fig. 5.22(d), one has

$$\frac{C}{R} = \frac{G}{1 + GH_2}$$

Backward substituting for the above equations, one obtains

$$\frac{C}{R} = \frac{G_1(G_2 + G_3)G_4 / [1 + H_1(G_2 + G_3)]}{1 + \frac{G_1 G_4 H_2 (G_2 + G_3)}{1 + H_1(G_2 + G_3)}}$$

Simplifying the above expression, one obtains

$$\frac{C}{R} = \frac{G_1(G_2 + G_3)G_4}{1 + H_1(G_2 + G_3) + G_1 G_4 H_2 (G_2 + G_3)}$$

This is the required closed-loop TF of the control system in Fig. 5.22(a).

5.4.4.4 Multiple Inputs

In some cases, a feedback control system may have several inputs, and it is necessary to determine the output when each input takes a particular form. Provided the system is linear, i. e., described by a linear differential equation, this can be achieved by the following.

Step 1: Set all but one input to zero.

Step 2: By rearranging the block diagram if necessary with the simplification method of the block diagram, determine the TF from the single nonzero input to the output.

Step 3: Repeat step 2 for all inputs.

Step 4: Add all TFs together to obtain the output to all inputs.

Example 5.11

Determine the output C of the two inputs R and D for the system shown in Fig. 5.23(a).

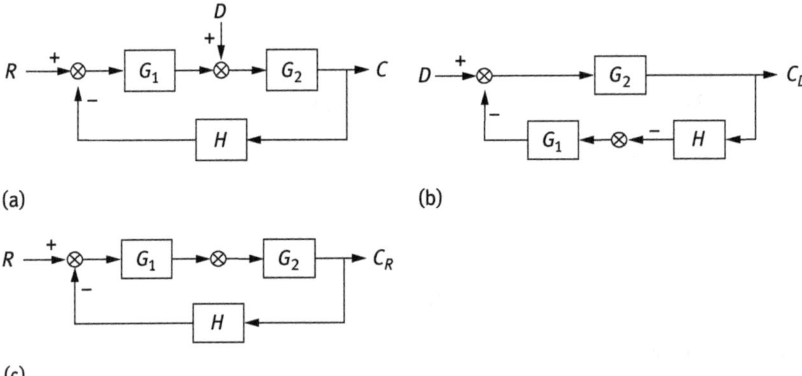

Fig. 5.23: Linear system with two inputs and its reduction of block diagram: (a) Original block diagram. (b) D as the input signal. (c) R as the input signal.

Solution

Setting R equal to zero and rearranging the block diagram produce the system shown in Fig. 5.24(b). The closed-loop TF becomes

$$\frac{C_D}{D} = \frac{G_2}{1 + G_1 G_2 H} \tag{5.27}$$

or

$$C_D = \frac{G_2}{1 + G_1 G_2 H} D \tag{5.28}$$

Now setting the other input D to zero produces the block diagram shown in Fig. 5.24(c). At this time, the TF becomes

$$\frac{C_R}{R} = \frac{G_1 G_2}{1 + G_1 G_2 H} \tag{5.29}$$

or

$$C_R = \frac{G_1 G_2}{1 + G_1 G_2 H} R \tag{5.30}$$

Hence, adding eqs. (5.28) and (5.30) results in

$$C = \frac{G_1 G_2}{1 + G_1 G_2 H} R + \frac{G_2}{1 + G_1 G_2 H} D \tag{5.31}$$

Generally, the input D is actually disturbance for the system output. Therefore, eq. (5.27) determines the relationship between the disturbance and the output and is called perturbation TF, while eq. (5.31) is called input–output TF. It's obvious that the two TFs have the same characteristic polynomial. However, they have different forward TFs.

Example 5.12

The closed-loop system is shown as Fig. 5.24(a). Find

(1) The TFs with the input $X(s)$ and the output $Y(s)$, $Y_1(s)$, $B(s)$, and $E(s)$, respectively.

(2) The TFs with the input $N(s)$ and the output $Y(s)$, $Y_1(s)$, $B(s)$, and $E(s)$, respectively.

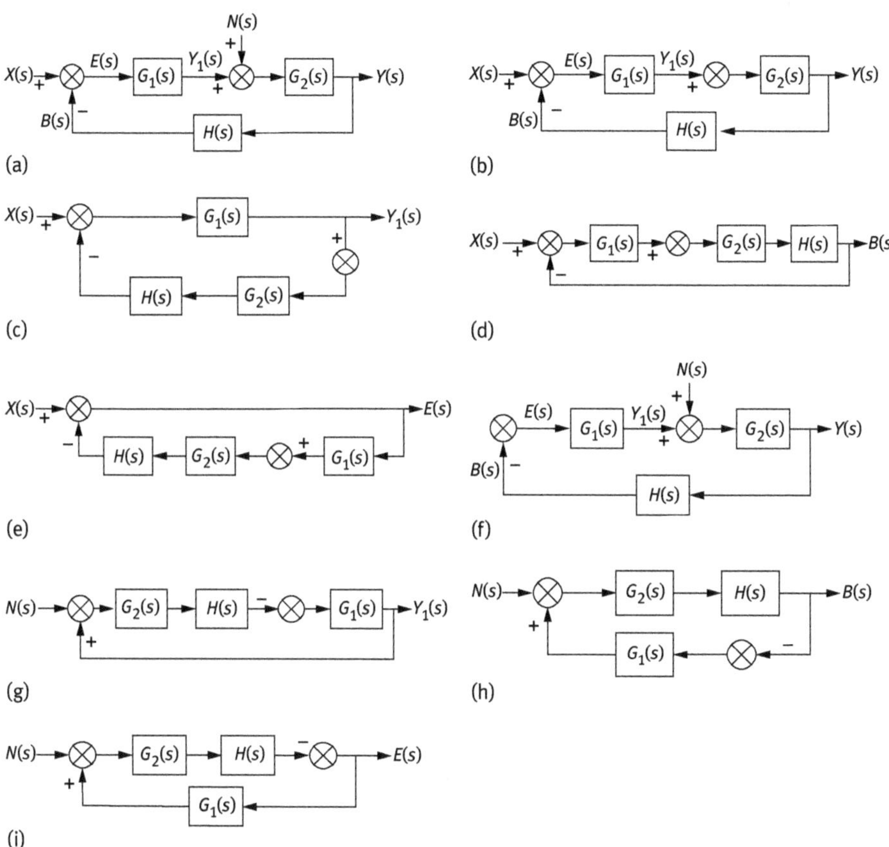

Fig. 5.24: Block diagrams for Example 5.12: (a) the closed-loop system for Example 5.12; (b) let the input of $N(s)$ equal to zero; (c) let $Y_1(s)$ output; (d) let $B(s)$ output; (e) let $E(s)$ output; (f) let the input signal $X(s)$ equal to zero; (g) let $Y_1(s)$ output; (h) let $B(s)$ output; and (i) let $E(s)$ output.

Solution (1)

By means of the same principle in Example 5.11, we first should let the input signal $N(s)$ equal zero, and the block diagram changes from Fig. 5.24(a) to (b).

(1) According to Fig. 5.24(b), the TF with the input $X(s)$ and the output $Y(s)$ is

$$G_Y(s) = \frac{Y(s)}{X(s)} = \frac{G_1(s)G_2(s)}{1 + G_1(s)G_2(s)H(s)}$$

(2) According to Fig. 5.24(c), the TF with the input $X(s)$ and the output $Y_1(s)$ is

$$G_{Y1}(s) = \frac{Y_1(s)}{X(s)} = \frac{G_1(s)}{1 + G_1(s)G_2(s)H(s)}$$

(3) According to Fig. 5.24(d), the TF with the input $X(s)$ and the output $B(s)$ is

$$G_B(s) = \frac{B(s)}{X(s)} = \frac{G_1(s)G_2(s)H(s)}{1 + G_1(s)G_2(s)H(s)}$$

(4) According to Fig. 5.24(e), the TF with the input $X(s)$ and the output $E(s)$ is

$$G_E(s) = \frac{E(s)}{X(s)} = \frac{1}{1 + G_1(s)G_2(s)H(s)}$$

Solution (2)

By means of the same principle in Example 5.11, first we should let the input signal $R(s)$ equal to zero, and the block diagram changes from Fig. 5.24(a) to Fig. (f).

(5) According to Fig. 5.24(f), the TF with the input $N(s)$ and the output $Y(s)$ is

$$G_Y(s) = \frac{Y(s)}{N(s)} = \frac{G_2(s)}{1 - G_1(s)G_2(s) \cdot [-H(s)]} = \frac{G_2(s)}{1 + G_1(s)G_2(s)H(s)}$$

(6) According to Fig. 5.24(g), the TF with the input $N(s)$ and the output $Y_1(s)$ is

$$G_{Y1}(s) = \frac{Y_1(s)}{N(s)} = \frac{-G_1(s)G_2(s)H(s)}{1 - G_2(s) \cdot [-H(s)] \cdot G_1(s)} = \frac{-G_1(s)G_2(s)H(s)}{1 + G_1(s)G_2(s)H(s)}$$

(7) According to Fig. 5.24(h), the TF with the input $N(s)$ and the output $B(s)$ is

$$G_B(s) = \frac{B(s)}{N(s)} = \frac{G_2(s)H(s)}{1 - G_2(s) \cdot [-H(s)] \cdot G_1(s)} = \frac{G_2(s)H(s)}{1 + G_1(s)G_2(s)H(s)}$$

(8) According to Fig. 5.24(i), the TF with the input $N(s)$ and the output $E(s)$ is

$$GY_1(s) = \frac{E(s)}{N(s)} = \frac{-G_2(s)H(s)}{1 + G_1(s)G_2(s)H(s)}$$

From the above eight TFs, we can get a conclusion: for the same closed-loop system, the TF of the system will be different because the different inputs and outputs result in different TFs of the forward path and different TFs of the feedback path. But anyway, for the same closed-loop system, the denominators of the TFs remain always the same because the denominator reflects the inherent characteristic of the system.

5.5 Plot the Block Diagrams

In this part, we will introduce a new method that can transfer the schematic diagram into the block diagram. According to the block diagram, one can find out the relationships and interaction among every links. Here are some useful tips about how to transfer the above two diagrams,

(1) Find the differential equations for every link of the system.
(2) Assume all initial conditions are equal to zero and take Laplace transform to get the TFs for every link.
(3) Plot sub-function block diagrams according to the TFs for every link.
(4) Combine all sub-block diagrams into a whole block diagram.

Example 5.13

Plot block diagram for the second-order *RC* passive circuit shown in Fig. 5.25, in which u_r is the input signal and u_c is the output signal for the whole system. R_1 and R_2 are the constants for resistors. C_1 and C_2 are the constants for capacitor. i_1 and i_2 are the current in the circuit. All initial conditions are zero.

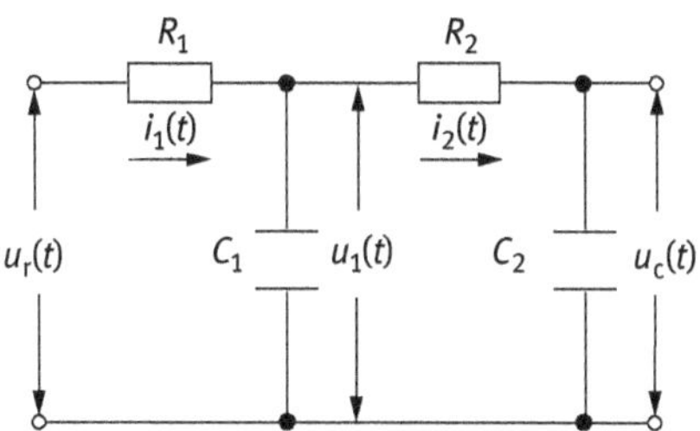

Fig. 5.25: The schematic diagram for the second-order *RC* passive circuit.

Solution

For the resistor R_1, the voltage between its two terminals is the input signal and the current through it is the output signal. According to the Ohm's law, we have

$$\frac{u_r(t) - u_1(t)}{i_1(t)} = R_1$$

Take Laplace transform of the above equation and rearrange it according to the relationship of the input and the output for the resistor R_1:

$$\frac{I_1(s)}{U_r(s) - U_1(s)} = \frac{1}{R_1}$$

In fact, it is a proportion link. The sub-function block diagram is shown in Fig. 5.26 (a).

For the capacitor C_1, the current through it is the input signal and the voltage between its two terminals is the output signal. According to the Ohm's law, we have

$$u_1(t) = \frac{1}{C_1}\int [i_1(t) - i_2(t)]dt$$

Take Laplace transform of the above equation and rearrange it according to the relationship of the input and the output for the resistor C_1:

$$\frac{U_1(s)}{I_1(s)-I_2(s)}=\frac{1}{C_1 s}$$

It is an integral link. The sub-function block diagram is shown in Fig. 5.26(b).

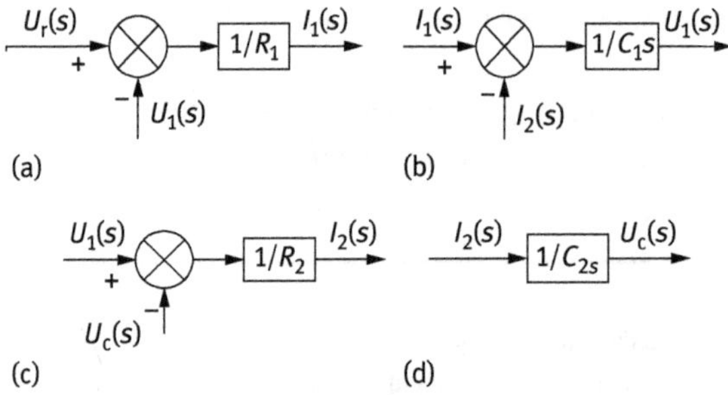

Fig. 5.26: The sub-function block diagrams for Example 5.13: (a) sub-function block diagram for R1 link. (b) sub-function block diagram for C1 link. (c) sub-function block diagram for R2 link. (d) sub-function block diagram for C2 link.

For the resistor R_2, the voltage between its two terminals is the input signal and the current through it is the output signal. According to the Ohm's law, we have

$$\frac{u_1(t)-u_c(t)}{i_2(t)}=R_2$$

Take Laplace transform of the above equation and rearrange it according to the relationship of the input and the output for the resistor R_2:

$$\frac{I_2(s)}{U_1(s)-U_c(s)}=\frac{1}{R_2}$$

It is a proportion link. The sub-function block diagram is shown in Fig. 5.26(c).

For the capacitor C_2, the current through it is the input signal and the voltage between its two terminals is the output signal. According to the Ohm's law, we have

$$u_c(t)=\frac{1}{C_2}\int i_2(t)\mathrm{d}t$$

Take Laplace transform of the above equation and rearrange it according to the relationship of the input and the output for the resistor C_2:

$$\frac{U_c}{I_2(s)}=\frac{1}{C_2 s}$$

It is an integral link. The sub-function block diagram is shown in Fig. 5.26(d).

Finally, according to sub-function block diagrams, together with input and output for the whole system, one can get the final block diagram, as shown in Fig. 5.27:

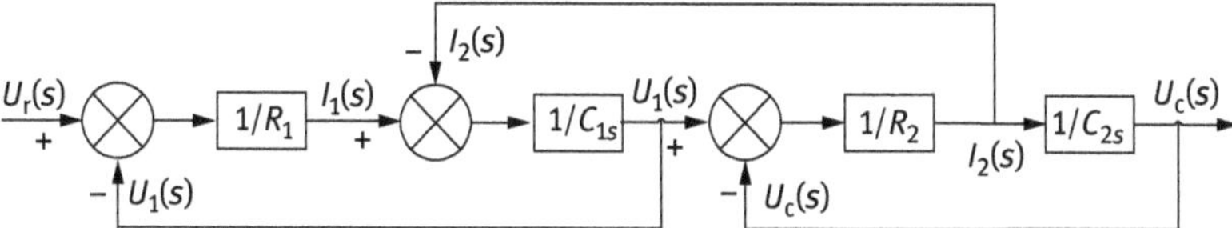

Fig. 5.27: The final block diagram for Example 5.13.

Example 5.14

Plot block diagram according to the given schematic diagram shown in Fig. 5.28 (a). The applied force $f_i(t)$ is the input variable and the displacement $x_o(t)$ is the output variable for the whole system. m_1 and m_2 are constants of masses. K_1, K_2, and C are all constants for springs and dashpot, respectively. Neglect the gravity for the masses. All initial conditions are zero.

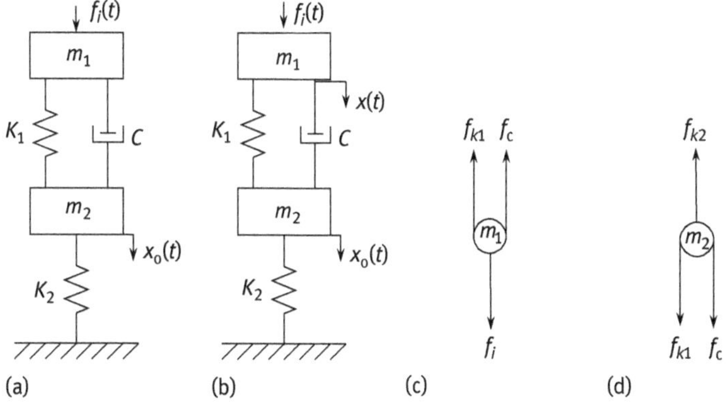

Fig. 5.28: The schematic diagram and free-body diagrams for Example 5.14: (a) the schematic diagram for mass–spring–damping system; (b) introduce an intermediate variable $x(t)$ to this system; (c) free-body diagram for mass 1; and (d) free-body diagram for mass 2.

Solution

First, introduce an intermediate variable $x(t)$ that is the displacement of mass 1, shown in Fig. 5.28(b). And then, in terms of the schematic diagram with the intermediate variable, one can make free-body diagrams for mass 1 and mass 2, respectively, shown in Fig. 5.28(c) and (d).

Write the differential equations for every element mass 1, spring 1, dashpot, mass 2, and spring 2:

$$m_1 \frac{d^2 x(t)}{dt^2} = f_i(t) - f_c(t) - f_{k_1}(t)$$

$$\begin{cases} f_{k_1}(t) = K_1 [x(t) - x_o(t)] \\ f_c(t) = C\left[\frac{dx(t)}{dt} - \frac{dx_o(t)}{dt}\right] \end{cases}$$

$$m_2 \frac{d^2 x_o(t)}{dt^2} = f_{k_1}(t) + f_c(t) - f_{k_2}(t)$$

$$f_{k_2}(t) = K_2 x_o(t)$$

Take Laplace transform of the above five differential equations in turns:

$$\frac{X(s)}{F_i(s) - F_c(s) - F_{k_1}(s)} = \frac{1}{m_1 s^2}$$

$$\begin{cases} \frac{F_{k_1}(s)}{X(s) - X_o(s)} = K_1 \\ \frac{F_c(s)}{X(s) - X_o(s)} = Cs \end{cases}$$

$$\frac{X_o(s)}{F_{k_1}(s) + F_c(s) - F_{k_2}(s)} = \frac{1}{m_2 s^2}$$

$$\frac{F_{k_2}(s)}{X_o(s)} = K_2$$

By means of Laplace transforms, one can plot every sub-block diagrams, shown in Fig. 5.29. Finally, combine the sub-block diagrams into the whole block diagram, as shown in Fig. 5.30.

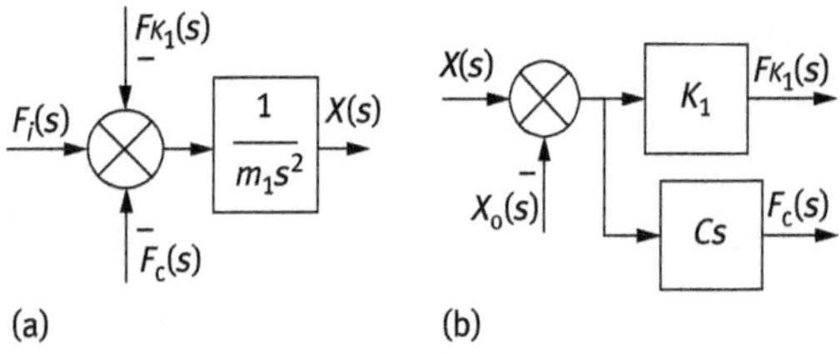

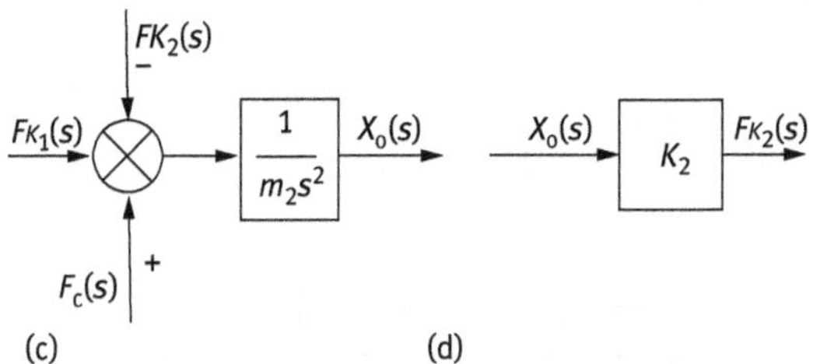

Fig. 5.29: The sub-function block diagrams for Example 5.14: (a) sub-function block diagram for m1 link. (b) sub-function block diagram for C link. (c) sub-function block diagram for m2 link. (d) sub-function block diagram for K2 link.

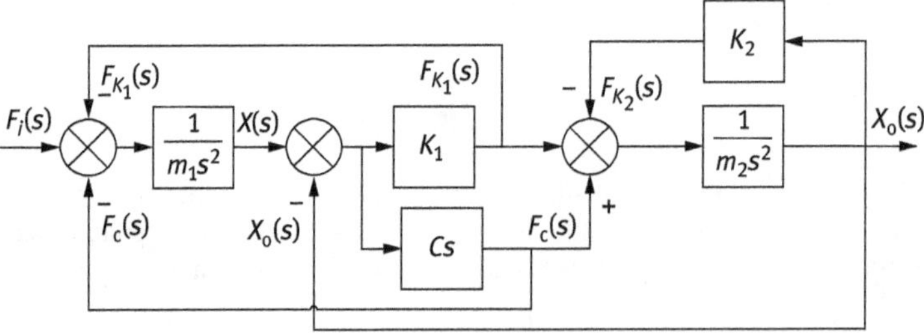

Fig. 5.30: The final block diagram for Example 5.14.

5.6 Signal Flow Diagrams

5.6.1 Introduction to Signal Flow Diagrams

Signal flow diagrams are primarily an alternative pictorial representation to block diagrams (see Fig. 5.31). All signals (variables) are represented by dots, called nodes, such as x_1, x_2, x_3, x_4, x_5, and related variables are joined by lines called directed branches. Each branch has an associated transmittance, which links node x_i to node x_j with zero transmittance from x_j to x_i, i.e., transmittance is unidirectional and is indicated by an arrow, such as a, b, c, and d.

An input variable is represented by a source node and an output (response) variable by a sink node. A path from a source node to a sink node, without passing through any other node more than once, is called a forward path, e.g., $x_1 \to x_2 \to x_3 \to x_4$ and $x_5 \to x_3 \to x_4$ in Fig. 5.31, where x_1 and x_5 are the source nodes and x_4 is the sink node. A closed (feedback) path is called a loop, e.g., $x_2 \to x_3 \to x_2$. The product of the transmittances of the branches forming a loop is called the loop transmittance, such as b and c.

Note the following points:

(1) The signal at a node is equal to the sum of all signals transmitted to the node, i.e., a node with more than one input is a summing point. Sometimes the transmittance may be negative.

(2) The transmittances are simply related to the TFs.

(3) The transmittances connected to the input/output nodes are both unity, and merely help to make the diagram clearer.

(4) The same rules of noninteraction as discussed in block diagram may apply to signal flow diagram.

5.6.2 Draw the Signal Flow Diagram

The source node in the signal flow diagram, $X(s)$, is the same one in the block, and the sink node in the signal flow diagram, $Y(s)$, is the same one in the block (see Fig. 5.32,). TF in the block is the transmittance $G(s)$ in the signal flow diagram.

About feedback loop, the corresponding relationship between block and signal flow diagram can be obtained from Fig. 5.33. The minus sign with $H(s)$ in the signal flow diagram represents the negative feedback in block.

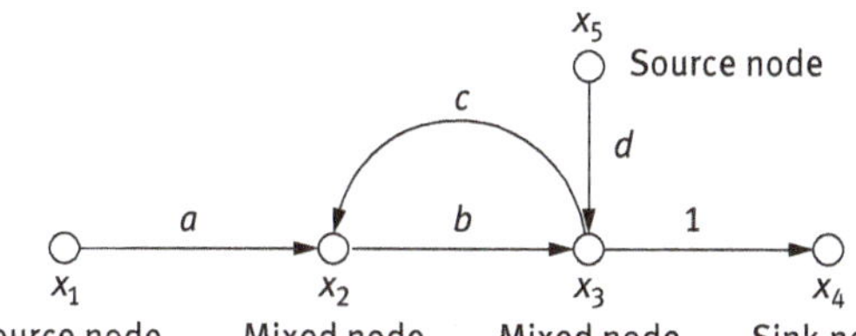

Fig. 5.31: Some elements of the signal flow diagram.

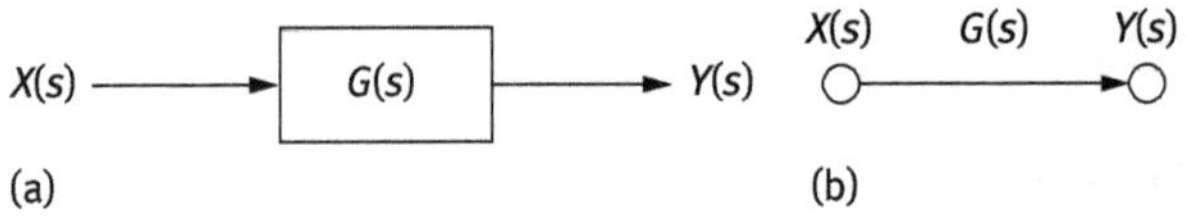

Fig. 5.32: Block diagram (a) and the corresponding signal flow diagram (b).

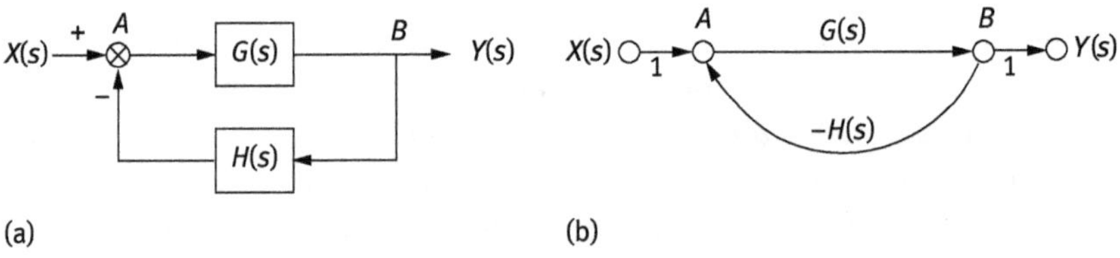

Fig. 5.33: Block diagram (a) and signal flow diagram (b) for feedback loop.

Example 5.15

Change the block in Fig. 5.34 into the signal flow diagram.

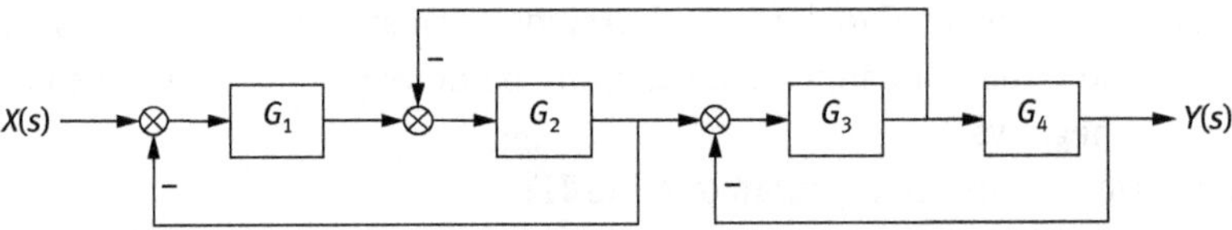

Fig. 5.34: Block for Example 5.15.

Solution

Figure 5.35 is the signal flow diagram changed for Fig. 5.34. Dotted lines in Fig. 5.35 are just a help for marking the corresponding points between block and signal flow diagram.

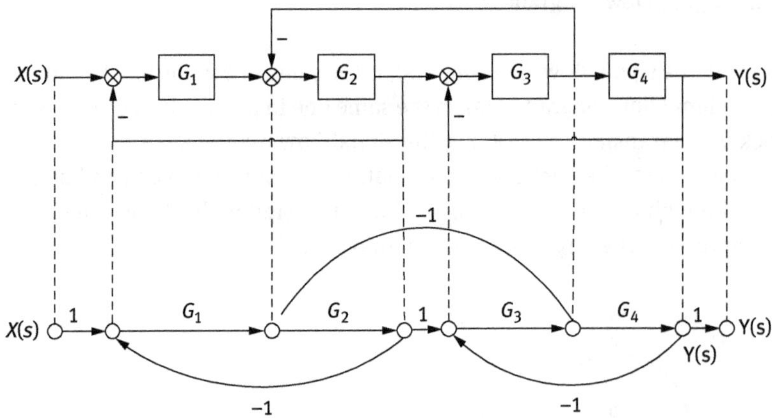

Fig. 5.35: Signal flow diagram for Fig. 5.34

Example 5.16
Change the block in Fig. 5.36 into the signal flow diagram.

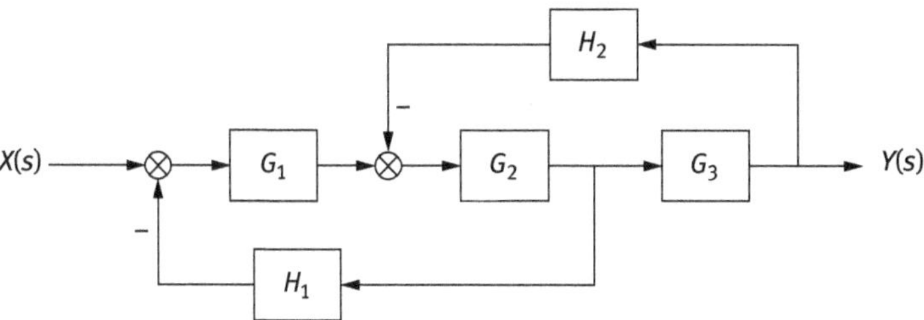

Fig. 5.36: Block for Example 5.16.

Solution
Figure 5.37 is the signal flow diagram changed for Fig. 5.36.

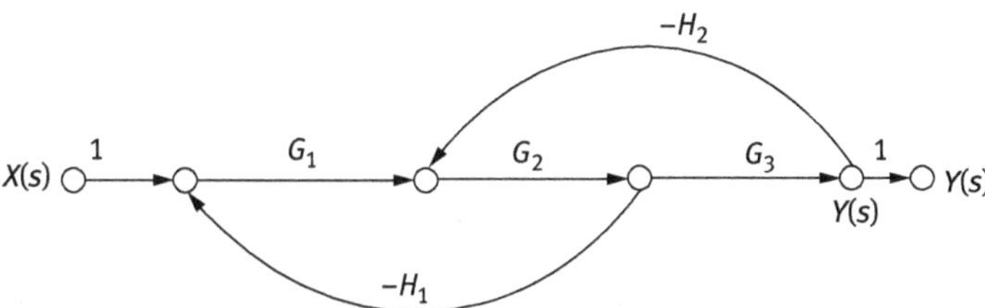

Fig. 5.37: Signal flow diagram for Fig. 5.36.

5.6.3 Mason's Gain Formula

Input–output relationships, for both block diagrams and signal flow diagrams, have been derived by reducing the complexity of the diagram step by step. Mason gives a formula to derive the input–output relation directly from the original signal flow diagram as an alternative to reduction.

Mason's gain formula states that net transmittance P, from a source node to a sink node, is given by

$$P = \frac{\sum_k P_k \cdot \Delta_k}{\Delta} \tag{5.32}$$

where k is the number of the open paths between the source node and sink node under consideration; P_k is the transmittance of the kth open path; $\Delta = 1 -$ (sum of all loop transmittances) + (sum of products of loop transmittances of all possible nontouching loops taken in pairs) – (sum of similar products taken three at a time + etc.; Δ_k is the value of Δ calculated for that part of the graph not touching the kth open path.

So, the cores of Mason's gain formula are as follows:

(1) All the forward path transmittances.
(2) All the loop transmittances.
(3) All the possible nontouching loop transmittances.
(4) Whether all the loop and all the forward path touch or not.

Example 5.17
Get the TF of the block in Fig. 5.34.

Solution
For Fig. 5.34,

(1) One forward path, P_1, and the forward path transmittance is $P_1 = G_1G_2G_3G_4$.
(2) Three loops, L_1, L_2, and L_3, the loop transmittances are $L_1 = -G_1G_2$, $L_2 = -G_3G_4$, and $L_3 = -G_2G_3$.
(3) Nontouching loop is L_1 and L_2, their product is $L_1\, L_2 = G_1G_2G_3G_4$.
(4) The three loops touch the only one forward path, hence $\Delta_1 = 1$.

So, the TF for Fig. 5.34 is

$$G(s) = \frac{P_1\Delta_1}{\Delta} = \frac{G_1G_2G_3G_4}{1 + G_1G_2 + G_3G_4 + G_2G_3 + G_1G_2G_3G_4}$$

Example 5.18
Get the TF of the signal flow diagram in Fig. 5.38.

Solution

(1) There are three forward paths, and the forward path transmittances are as follows: $P_1 = G_1G_2G_3G_4G_5$, $P_2 = G_1G_6G_4G_5$, $P_3 = G_1G_2G_7$.
(2) There are four feedback loops, and the feedback loop transmittances are as follows: $L_1 = -G_4H_1$, $L_2 = -G_2G_3G_4G_5H_2$, $L_3 = -G_6G_4G_5H_2$, $L_4 = -G_2G_7H_2$.
(3) Nontouching feedback loops are L_1 and L_4; hence, $\Delta = 1 - (L_1 + L_2 + L_3 + L_4) + L_1L_4$.
(4) All feedback loops touch the forward path P_1; hence, $\Delta_1 = 1$.
(5) All feedback loops touch the forward path P_2; hence, $\Delta_2 = 1$.

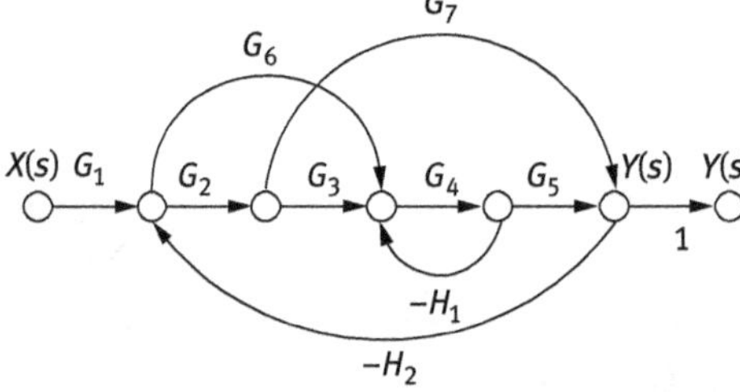

Fig. 5.38: Signal flow diagram for Example 5.18.

(6) The feedback loop L_1 does not touch the forward path P_3; hence, $\Delta_3 = 1 - L_1$.

So, the TF for Fig. 5.38 is

$$G(s) = \frac{P_1\Delta_1 + P_2\Delta_2 + P_3\Delta_3}{\Delta} = \frac{G_1G_2G_3G_4G_5 + G_1G_6G_4G_5 + G_1G_2G_7(1 + G_4H_1)}{1 + G_4H_1 + G_2G_7H_2 + G_4G_6G_5H_2 + G_2G_3G_4G_5H_2 + G_2G_7G_4H_1H_2}$$

5.7 Problems

P5.1. Reduce the following system shown in Fig. 5.39 to one block and in so doing derive the closed-loop TF C/R.

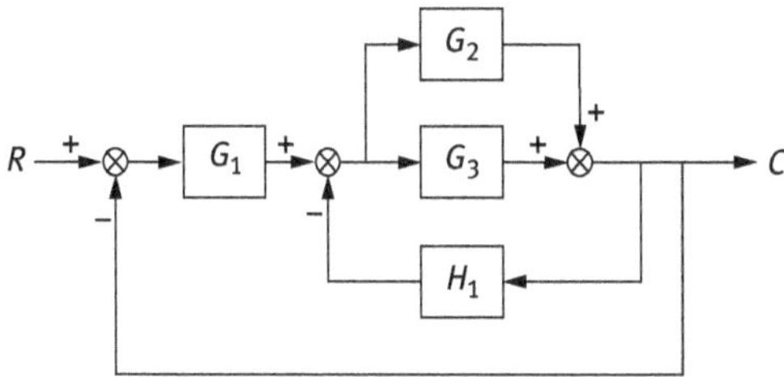

Fig. 5.39: An instance for reduction of block diagram.

P5.2. Simplify the block diagram shown in Fig. 5.40 and determine the closed-loop TF C/R.

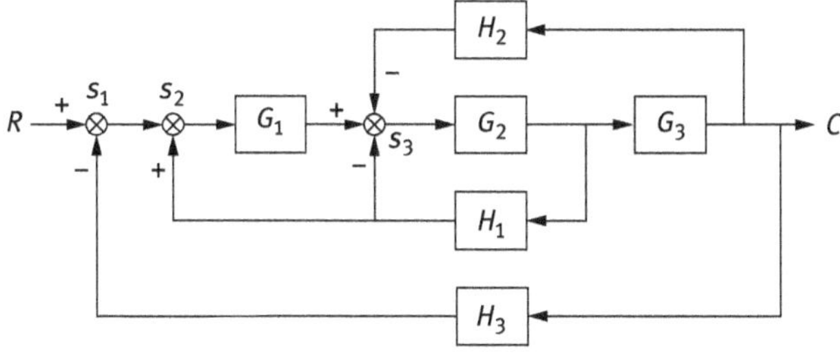

Fig. 5.40: An example to reduce the block diagram.

P5.3. Get the TF of the block in Fig. 5.41.

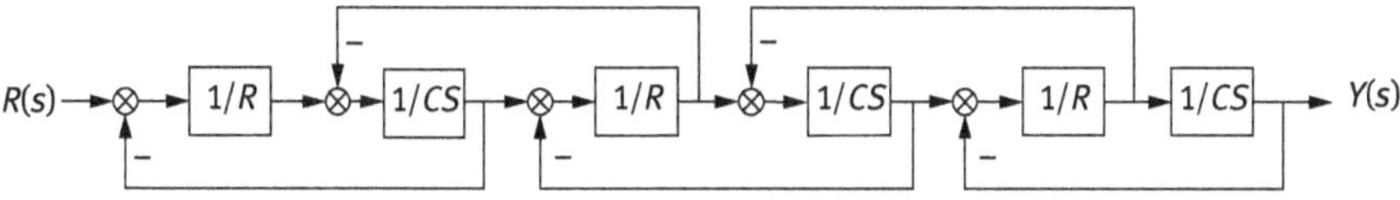

Fig. 5.41: Transfer function block for a circuit system.

P5.4. Fig. 5.42 shows a feedback control system, in which $N(s)$ is the disturbance, $R(s)$ is the input, and $C(s)$ is the output, respectively.

(1) Find the TF $C(s)/R(s)$.

(2) Find the TF $C(s)/N(s)$.

(3) If we want to eliminate the influence that the disturbance $N(s)$ brings to this system, which means, $C(s)/N(s) = 0$, please find the expression of $G_0(s)$.

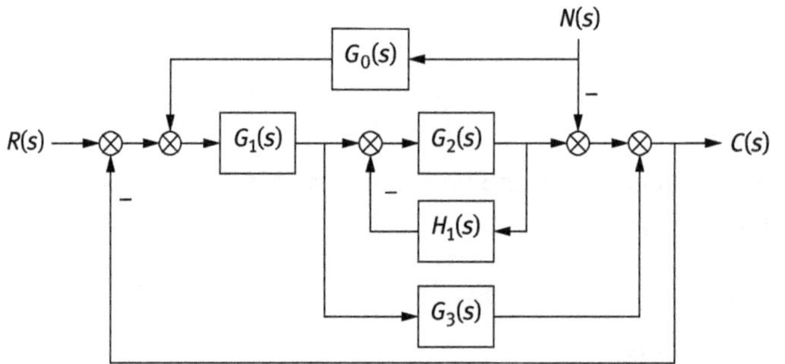

Fig. 5.42: Block diagram.

P5.5. Find the TFs of the systems shown in Fig. 5.43 with Mason's gain formula.

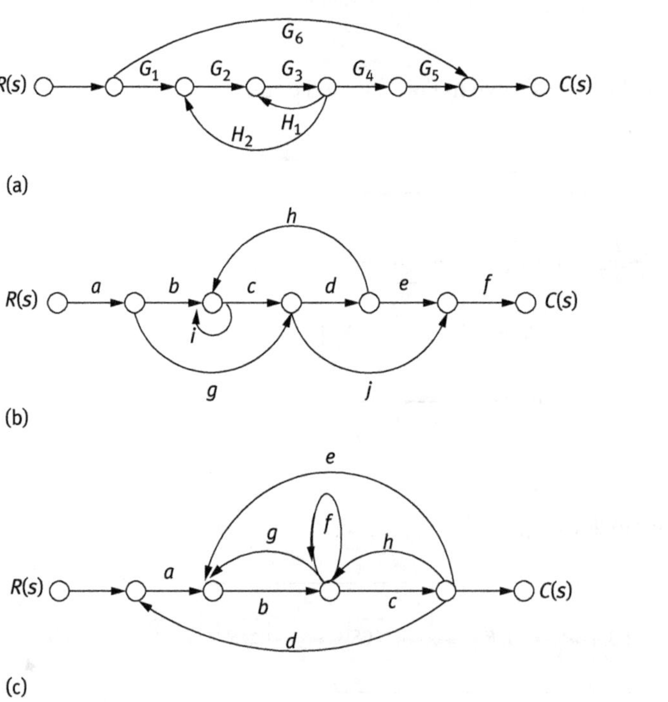

Fig. 5.43: Signal flow diagrams for P5.5.

P5.6. The schematic diagram of the system is shown in Fig. 5.44. k_1, k_2, f, m_1, and m_2 are all constants. $f_1(t)$ is the applied force and $x_2(t)$ is the output variable of the displacement.

(1) Plot the free-body diagrams for m_1 and m_2, respectively.

(2) Find the TF with the input variable of the applied force $f_1(t)$ and the output variable of the displacement $x_2(t)$. (Neglect the gravity.)

(3) Plot block diagram according to the given schematic diagram.

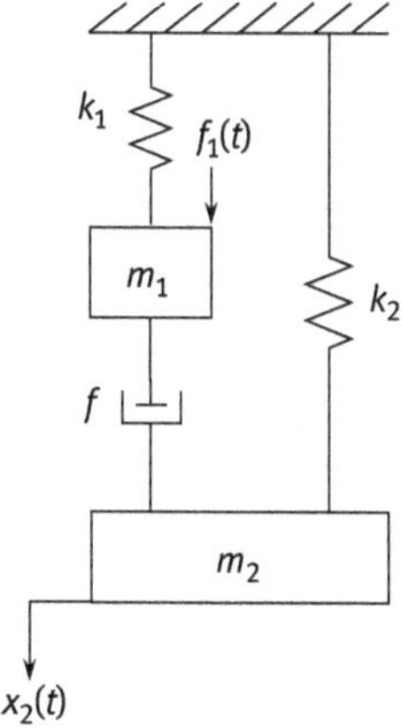

Fig. 5.44: The schematic diagram for P5.6.

6 Time Response Analysis of Control Systems

6.1 Introduction

There are two parts of the system analysis: the fundamentals for the system analysis and the method for the system analysis. The former, in fact, consists of two processes: confirming the mathematical model for the system and confirming the transfer function (TF) for the system. The latter consists of two analysis methods: the response analysis method in time domain and the response analysis method in frequency domain. In the previous chapters, especially in Chapter 5, we have learnt about the fundamental theory. So, in Chapters 6 and 7, we will introduce some principles of the response analysis methods in time domain and and frequency domain, respectively.

For a typical input signal, the output $y(t)$ of the control system will change with time in time domain. The function that depicts the change is called the time response of the control system. In fact, the solution for the differential equation of the control system is the time response function. The time response consists of the transient response and the steady-state response. The transient response is a response process in which the output variable changes from the initial state to the steady state when the system is subjected to some kind of input signal. The steady-state response is the output value that we get when time tends to infinity when the system is subjected to some kind of input signal. The difference between the steady-state response and the desired output value can be used to measure the accuracy of the system.

Figure 6.1 shows the time response curve for the system subjected to the unit step input signal. The output of the system reaches the steady state when time is the settling time t_s. The response process during 0 to t_s is called the transient response. If the output converges to some steady-state value as time tends to infinity, the system is a stable one. And at this time, the output of the system is the steady-state response $y(\infty)$. If the curve $y(t)$ oscillates in equal amplitude or in divergent amplitude, the system is an unstable one. Transient response reflects the dynamic characteristics of the system.

6.2 Time Response from TF

6.2.1 Response of First-Order System

A physical system that is described by a first-order differential equation is identified as a first-order system, such as the mechanical first-order system shown in Fig. 6.2. The input variable x_i and output variable x_o are related by the following equation obtained by equating the forces at point A,

https://doi.org/10.1515/9783110573275-006

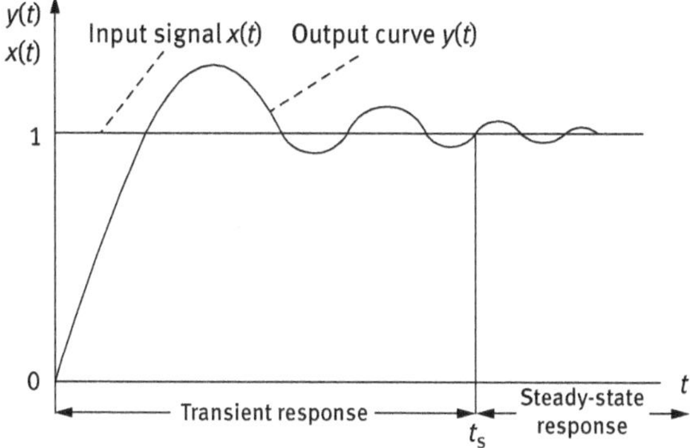

Fig. 6.1: Time response curve for the system subjected to the unit step input signal.

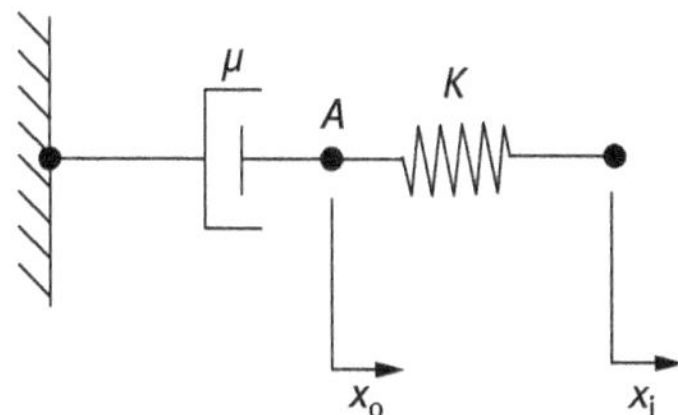

Fig. 6.2: A mechanical first-order system.

$$K(x_{\mathrm{i}} - x_{\mathrm{o}}) = \mu \dot{x}_{\mathrm{o}}$$

Rearranging the above equation,

$$\frac{\mu}{K}\dot{x}_{\mathrm{o}} + x_{\mathrm{o}} = x_{\mathrm{i}}$$

Taking the Laplace transform of it and rearranging,

$$\frac{X_{\mathrm{o}}}{X_{\mathrm{i}}} = \frac{1}{1 + (\mu/K)s} = \frac{1}{1 + \tau s}$$

where τ is introduced to represent the time constant for this first-order mechanical system.

Another example of the first-order system is an electrical first-order system, shown in Fig. 6.3. Applying Kirchhoff's Law and Ohm's Law to this circuit,

$$u_{\mathrm{i}} - u_{\mathrm{o}} = CR\frac{\mathrm{d}u_{\mathrm{o}}}{\mathrm{d}t}$$

Taking the Laplace transform of it and rearranging,

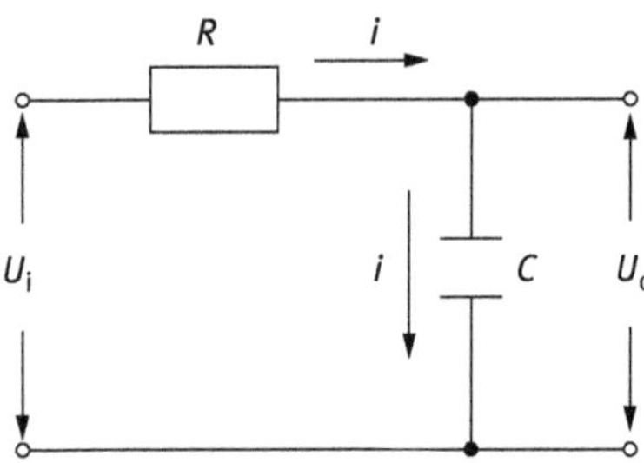

Fig. 6.3: An electrical first-order system.

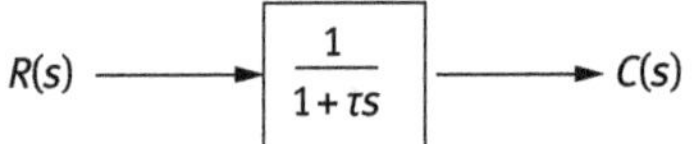

Fig. 6.4: The function block diagram for the first-order system.

$$\frac{U_o}{U_i} = \frac{1}{1+RCs} = \frac{1}{1+\tau s}$$

where τ is the time constant of the first-order electrical system.

From the previous two examples, for some systems represented in the first-order form, we can get the function diagram shown in Fig. 6.4. And the generalized TF between the input $R(s)$ and the output $C(s)$ of these systems may be related by eq. (6.1)

$$\frac{C(s)}{R(s)} = \frac{1}{1+\tau s} \tag{6.1}$$

Now, we will introduce different input signals into this first-order system.

6.2.1.1 Unit Step Response

For the unit step signal $r(t) = 1$, the Laplace transform is

$$R(s) = \frac{1}{s}$$

Substituting the above equation into eq. (6.1), we get the output signal $C(s)$,

$$C(s) = \frac{1}{s(1+\tau s)} = \frac{1}{s} - \frac{\tau}{1+\tau s} = \frac{1}{s} - \frac{1}{s+\frac{1}{\tau}}$$

Taking the Laplace transform inversion, we get the time response of the first-order system as

$$c(t) = 1 - e^{-t/\tau} \tag{6.2}$$

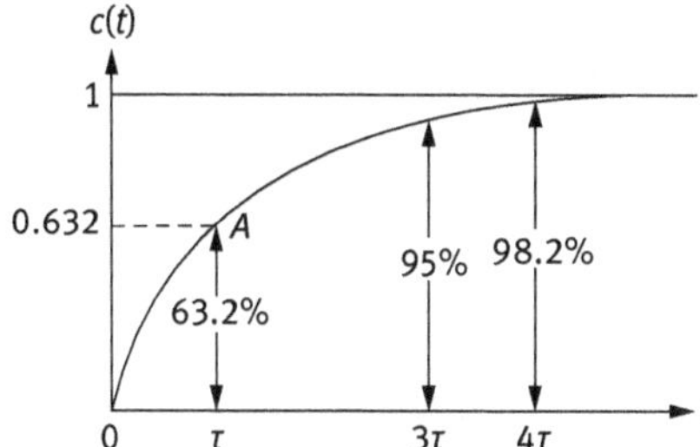

Fig. 6.5: Step response.

From eq. (6.2), one can get the unit step response curve of the first-order system, shown in Fig. 6.5. When $t = \tau$, one gets $c(\tau) = 0.632$, where τ is the time constant of this first-order system. It is an important index for system analysis. When $t = 3\tau$, we get $c(3\tau) = 0.95$. When $t = 4\tau$, we get $c(3\tau) = 0.982$. The previous evaluation reveals that when $t \in (3\tau, 4\tau)$, the error of the system lies in (2%, 5%), which could meet the design requirement of the system. Thus, the time of 3τ or 4τ is known as settling time of the system.

6.2.1.2 Unit Ramp Response

For the unit ramp signal $r(t) = t$, the Laplace transform is

$$R(s) = \frac{1}{s^2}$$

Substituting the aforementioned equation into eq. (6.1), we get the output signal $C(s)$,

$$C(s) = \frac{1}{s^2(1+\tau s)} = \frac{1}{s^2} - \frac{\tau}{s} + \frac{\tau}{s+1/\tau}$$

And then, taking the Laplace transform, inversion, we get the time response of the first-order system,

$$c(t) = t - \tau + \tau e^{-t/\tau} = t - \tau(1 - e^{-t/\tau}) \tag{6.3}$$

From eq. (6.3), one can get the unit ramp response curve of the first-order system, shown in Fig. 6.6. When time tends to infinity, the error equals to τ. Hence, in this case, the output could follow the ramp input signal with the error of τ.

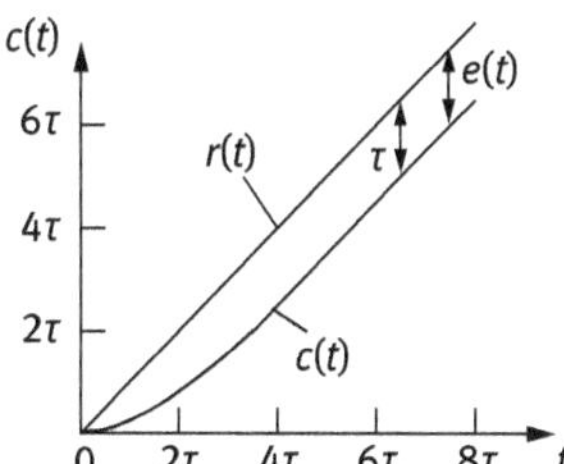

Fig. 6.6: Ramp response.

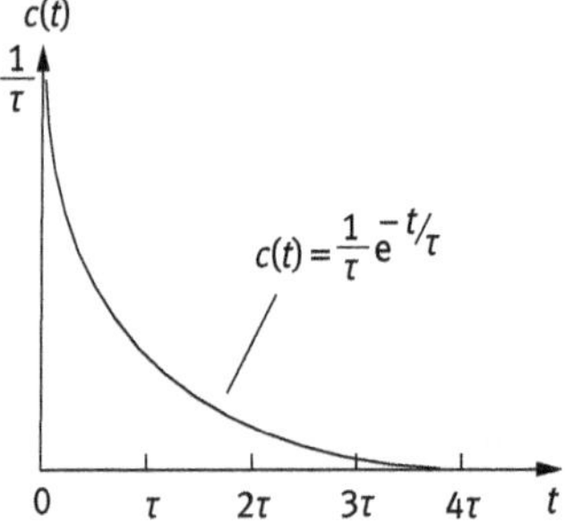

Fig. 6.7: Impulse response.

6.2.1.3 Unit Impulse Response

For the unit impluse signal $\sigma(t)$, the Laplace transform is

$$R(s) = 1$$

Substituting the above equation into eq. (6.1), we get the output signal $C(s)$,

$$C(s) = \frac{1}{\tau s + 1}$$

Taking the Laplace transform inversion, we get the time response of the first-order system,

$$c(t) = \frac{1}{\tau} \mathrm{e}^{-t/\tau} \tag{6.4}$$

From eq. (6.4), one can get the unit impluse response curve of the first-order system, shown in Fig. 6.7.

6.2.1.4 Remarks

If (a) the input signal is not a unit signal, and at the same time, (b) the numerator of the first-order system TF is k, known as gain, not 1, that is

(1) Step input signal, $R(s) = \frac{r}{s}$.
(2) Ramp input signal, $R(s) = \frac{r}{s^2}$.
(3) Impulse input signal, $R(s) = r$.
(4) First-order system TF, $\frac{C(s)}{R(s)} = \frac{k}{1+\tau s}$.

Thus, time responses are as follows:

(1) Step response, $c(t) = k \cdot r(1 - \mathrm{e}^{-t/\tau})$.
(2) Ramp response, $c(t) = k \cdot r(t - \tau + \tau \mathrm{e}^{-t/\tau})$.
(3) Impulse response, $c(t) = \frac{k \cdot r}{\tau} \mathrm{e}^{-t/\tau}$.

Example 6.1a

The open-loop TF of a unit negative feedback system is

$$G(s) = \frac{20}{0.21s + 1}$$

Find time response subjected to a unit step input signal.

Solution

First, we find the closed-loop TF for this system,

$$\Phi(s) = \frac{Y(s)}{X(s)} = \frac{G(s)}{1 + G(s)} = \frac{20}{0.21s + 21} = \frac{0.95}{0.01s + 1}$$

From the aforementioned equation, the closed-loop TF is also a first-order system. At the same time, $k = 0.95$, $\tau = 0.01$, and $r = 1$. So, the time response subjected to a unit step input is

$$c(t) = k \cdot r \cdot \left(1 - e^{-t/\tau}\right) = 0.95 \cdot \left(1 - e^{-t/0.01}\right)$$

Example 6.1b

The open-loop TF of a unit negative feedback system is

$$G(s) = \frac{20}{0.21s + 1}$$

Find time response subjected to a step input $X(s) = 0.8/s$.

Solution

First, we find the closed-loop TF for the system,

$$\Phi(s) = \frac{Y(s)}{X(s)} = \frac{G(s)}{1 + G(s)} = \frac{20}{0.21s + 21} = \frac{0.95}{0.01s + 1}$$

From the above equation, the closed-loop TF is also a first-order system. In this case, $k = 0.95$, $\tau = 0.01$, and $r = 0.8$. Thus, its time response subjected to a step input is

$$c(t) = k \cdot r \cdot \left(1 - e^{-t/\tau}\right) = 0.76 \cdot \left(1 - e^{-t/0.01}\right)$$

Summary from above two examples

The transient response is the time response for the whole system. Hence, for the closed-loop system, when we want to analyze the transient response, we should first get the closed-loop TF of the system according to the open-loop TF.

Example 6.2

The differential equation of the control system is $2.5\mathrm{d}y(t)/\mathrm{d}t + y(t) = 20x(t)$. Evaluate the unit impulse response $g(t)$ and the unit step response $h(t)$. (All initial condition terms are zero.)

Solution

Referring to the differential equation together with the Laplace transform and the initial conditions, we can get the Laplace transform and the TF of this system as

$$2.5sY(s) + Y(s) = 20X(s)$$
$$\Phi(s) = \frac{Y(s)}{X(s)} = \frac{20}{1+2.5s}$$

where $k = 20$ and $\tau = 2.5$. The unit impulse response is

$$y(t) = \frac{k \cdot r}{\tau} e^{-t/\tau} = \frac{20 \times 1}{2.5} \cdot e^{-t/2.5} = 8e^{-0.4t}$$

The unit step response is

$$y(t) = k \cdot r \cdot \left(1 - e^{-t/\tau}\right) = 20 \times 1 \times \left(1 - e^{-t/2.5}\right) = 20\left(1 - e^{-0.4t}\right)$$

6.2.2 Response of Second-Order System

Here is a passive electrical circuit that is introduced to explain the response of the second-order system, as shown in Fig. 6.8.

In Fig. 6.8, the input variable is the applied voltage u_i and the output variable is the voltage across the capacitor u_o. According to Kirchhoff's Law and Ohm's Law,

$$\begin{cases} u_i = u_R + u_L + u_C = Ri + L\frac{di}{dt} + \frac{1}{C}\int_0^t i dt & (6.5a) \\ u_o = u_C = \frac{1}{C}\int_0^t i dt & (6.5b) \end{cases}$$

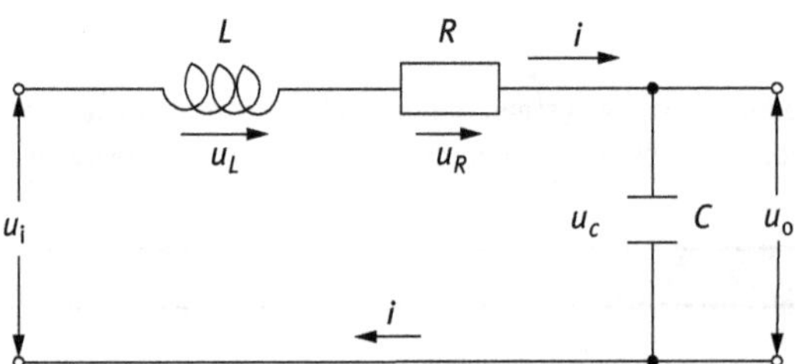

Fig. 6.8: A passive electrical circuit.

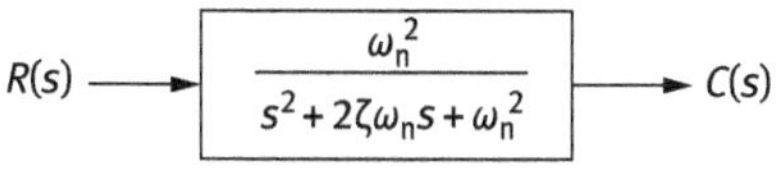

$$R(s) \longrightarrow \boxed{\frac{\omega_n^2}{s^2+2\zeta\omega_n s+\omega_n^2}} \longrightarrow C(s)$$

Fig. 6.9: The function block diagram for the generalized second-order system.

Taking the Laplace transform under the zero initial conditions,

$$\begin{cases} (Ls+R+\frac{1}{Cs})I=U_i & (6.5c) \\ \frac{1}{Cs}\cdot I=U_o \rightarrow I=U_o\cdot Cs & (6.5d) \end{cases}$$

Substituting eq. (6.5d) into eq. (6.5c) to eliminate the variable I yields the TF,

$$\frac{U_o}{U_i}=\frac{1}{LCs^2+RCs+1} \tag{6.5e}$$

Based on the above analysis, the general equation for second-order system is given by

$$\frac{C}{R}=\frac{\omega_n^2}{s^2+2\zeta\omega_n s+\omega_n^2} \tag{6.6}$$

where ω_n is the undamped natural frequency and ζ is the damping ratio. The function block diagram for the generalized second-order system is shown in Fig. 6.9.

Comparing the TF (e) with the general form of the second-order system,

$$\omega_n=\sqrt{\tfrac{1}{LC}},\quad \zeta=\tfrac{R}{2}\sqrt{\tfrac{C}{L}}$$

Considerable information can be obtained regarding the nature of a second-order system by studying its behavior when subjected to a step input. Substituting the unit step input into eq. (6.6) and using the partial fraction method, we get

$$C(s)=\frac{\omega_n^2}{s(s^2+2\zeta\omega_n s+\omega_n^2)}=\frac{1}{s}-\frac{s+2\zeta\omega_n}{(s+\zeta\omega_n)^2+\omega_n^2(1-\zeta^2)}$$

We introduce a new terminology, the characteristic equation of second-order system, which is the equation of the denominator of eq. (6.6),

$$s^2+2\zeta\omega_n s+\omega_n^2=0 \tag{6.7}$$

And the two roots of the characteristic equation are

$$s_1,s_2=-\zeta\omega_n\pm\omega_n\sqrt{\zeta^2-1} \tag{6.8}$$

The classification of the system with respect to the values of ζ is given as follows:

(1) Undamped case ($\zeta = 0$): $s_1, s_2 = \pm j\omega_n$.

(2) Underdamped case ($0 < \zeta < 1$): $s_1, s_2 = -\zeta\omega_n \pm j\omega_n\sqrt{1-\zeta^2}$.

(3) Critical case ($\zeta = 1$): $s_1, s_2 = -\omega_n$.

(4) Overdamped case ($\zeta > 1$): $s_1, s_2 = -\zeta\omega_n \pm \omega_n\sqrt{\zeta^2-1}$.

In Table 6.1, we analyze the aforementioned four cases one by one in detail.

6.2.2.1 $\zeta = 0$

For the undamped response ($\zeta = 0$), the roots are

$$s_1, s_2 = -\zeta\omega_n \pm j\omega_n\sqrt{1-\zeta^2} = \pm j\omega_n$$

Table 6.1: Second-order response as a function of damping cases.

ζ	Characteristic roots	Poles distribution	Step response
$\zeta = 0$	$s_1, s_2 = \pm j\omega_n$	$j\omega$; s-plane; $j\omega_n$; σ; $-j\omega_n$	$c(t)$; 1; 0; t; Undamped
$0 < \zeta < 1$	$s_1, s_2 = -\zeta\omega_n \pm j\omega_n\sqrt{\zeta^2-1}$	$j\omega$; s-plane; $j\omega_n\sqrt{\zeta^2-1}$; $-\zeta\omega_n$; σ; $-j\omega_n\sqrt{\zeta^2-1}$	$c(t)$; 1; 0; t; Underdamped
$\zeta = 1$	$s_1, s_2 = -\omega_n$	$j\omega$; s-plane; σ; $-\omega_n$	$c(t)$; 1; 0; Critically damped; t
$\zeta > 1$	$s_1, s_2 = -\zeta\omega_n \pm \omega_n\sqrt{\zeta^2-1}$	$j\omega$; s-plane; $-\zeta\omega_n + \omega_n\sqrt{\zeta^2-1}$; σ; $-\zeta\omega_n - \omega_n\sqrt{\zeta^2-1}$	$c(t)$; 1; 0; Overdamped; t

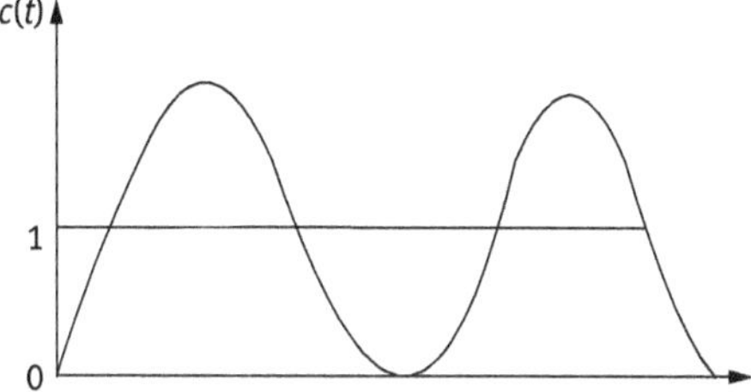

Fig. 6.10: Second-order undamped response.

Substituting the unit step input into the eq. (6.6), the output is

$$\begin{aligned} C(s) &= \frac{\omega_n^2}{s\left(s+\zeta\omega_n-\omega_n\sqrt{\zeta^2-1}\right)\left(s+\zeta\omega_n+\omega_n\sqrt{\zeta^2-1}\right)} \\ &= \frac{\omega_n^2}{s(s-\omega_n j)(s+\omega_n j)} \\ &= \frac{1}{s}+\frac{-s}{s^2+\omega_n^2} \end{aligned}$$

Thus, the Laplace transform inversion is

$$c(t) = 1 - \cos\omega_n t$$

The second-order undamped response curve under the step input signal is shown in Fig. 6.10. From Fig. 6.10, the output response is an equal amplitude cosine curve. In the classical control system, equal amplitude oscillation is treated as a critical and unstable situation.

6.2.2.2 $0 < \zeta < 1$

In the case of the underdamped response ($0 < \zeta < 1$), the roots are as shown in Fig. 6.11,

$$s_{1,2} = -\zeta\omega_n \pm \omega_n\sqrt{1-\zeta^2}\cdot j = -\zeta\omega_n \pm \omega_d \cdot j$$

where ω_d is called the damped natural frequency,

$$\omega_d = \omega_n\sqrt{1-\zeta^2} \tag{6.9}$$

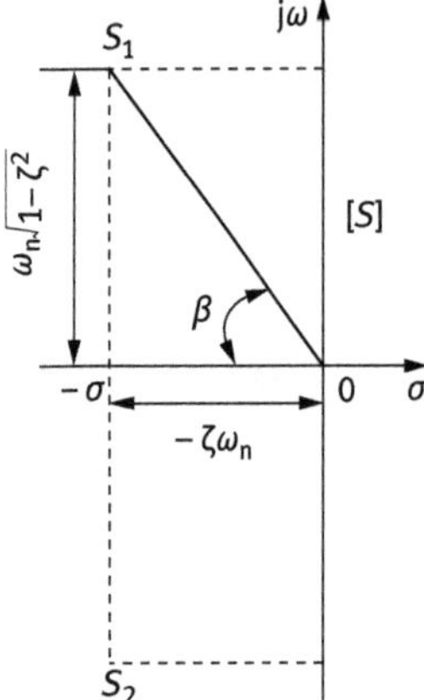

Fig. 6.11: The two roots distribution in the case of underdamped response ($0 < \zeta < 1$).

Substituting the unit step input into eq. (6.6), the output is

$$C(s) = \frac{1}{s} - \frac{s + 2\zeta\omega_n}{(s+\zeta\omega_n)^2 + \omega_n^2(1-\zeta^2)}$$

Factorizing the above equation,

$$C(s) = \frac{1}{s} - \frac{s + \zeta\omega_n}{(s+\zeta\omega_n)^2 + \omega_n^2(1-\zeta^2)} - \frac{\zeta\omega_n}{(s+\zeta\omega_n)^2 + \omega_n^2(1-\zeta^2)}$$

Substituting ω_d into it,

$$C(s) = \frac{1}{s} - \frac{s + \zeta\omega_n}{(s+\zeta\omega_n)^2 + \omega_d^2} - \frac{\zeta\omega_n}{(s+\zeta\omega_n)^2 + \omega_d^2}$$

Taking the Laplace transform inversion,

$$c(t) = 1 - e^{-\zeta\omega_n t}\left(\cos\omega_d t + \frac{\zeta}{\sqrt{1-\zeta^2}}\sin\omega_d t\right)$$
$$= 1 - \frac{e^{-\zeta\omega_n t}}{\sqrt{1-\zeta^2}}\sin\left(\omega_d t + \arctan\frac{\sqrt{1-\zeta^2}}{\zeta}\right)$$

Refer to Fig. 6.11 and let

$$\beta = \arctan\left(\frac{\sqrt{1-\zeta^2}}{\zeta}\right) \tag{6.10}$$

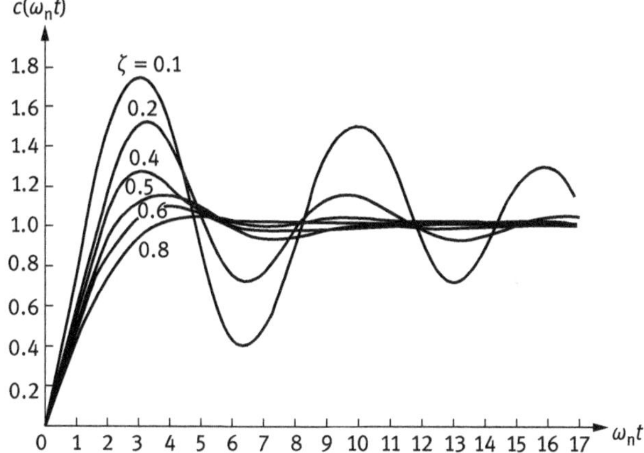

Fig. 6.12: Second-order underdamped responses.

Substituting eq. (6.10) into the equation of $c(t)$,

$$c(t) = 1 - \frac{e^{-\zeta\omega_n t}}{\sqrt{1-\zeta^2}}\sin(\omega_d t + \beta)$$

Figure 6.12 shows the response curves under the different damping ratio when $0 < \zeta < 1$, the output response curve will be convergent. The smaller the value of ζ, the more serious is the amplitude of oscillation. In general, we will choose the value of the interval (0.1, 0.8) because when ζ lies in this interval, the system will work at the suitable situation with comprehensive consideration of stability and rapidity. Especially, $\zeta = 0.707$ is known as the best perfect damping ratio.

6.2.2.3 $\zeta = 1$

In the case of the critically damped response, that is, $\zeta = 1$, the roots are

$$s_1, s_2 = -\zeta\omega_n \pm j\omega_n\sqrt{\zeta^2 - 1} = -\omega_n$$

Substituting the unit step input into eq. (6.6), the output is

$$C(s) = \frac{\omega_n^2}{s\left(s + \zeta\omega_n - \omega_n\sqrt{\zeta^2-1}\right)\left(s + \zeta\omega_n + \omega_n\sqrt{\zeta^2-1}\right)}$$

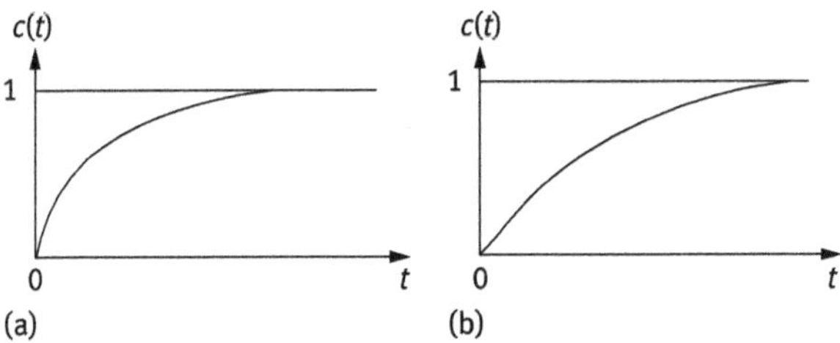

Fig. 6.13: Second-order response when $\zeta \geq 1$: (a) critically damped response and (b) overdamped response.

After simplifying with $\zeta = 1$,

$$C(s) = \frac{\omega_n^2}{s(s+\omega_n)^2}$$

By factorizing, we get

$$C(s) = \frac{1}{s} - \frac{1}{s+\omega_n} - \frac{\omega_n}{(s+\omega_n)^2}$$

Thus, according to the inverse Laplace transform, we have

$$c(t) = 1 - e^{-\omega_n t} - \omega_n t e^{-\omega_n t} = 1 - e^{-\omega_n t}(1+\omega_n t)$$

The response curve is shown in Fig. 6.13(a). According to Fig. 6.13(a), you will find that output response curve will come to the input signal when time tends to infinity. In fact, error must exist between the output and input for the application control system.

6.2.2.4 $\zeta > 0$

For the overdamped response, that is, $\zeta > 1$, the roots are

$$s_1, s_2 = -\zeta\omega_n \pm \omega_n\sqrt{\zeta^2 - 1}$$

Substituting the unit step input into eq. (6.6), the output is

$$\begin{aligned} C(s) &= \frac{\omega_n^2}{s\left(s+\zeta\omega_n - \omega_n\sqrt{\zeta^2-1}\right)\left(s+\zeta\omega_n + \omega_n\sqrt{\zeta^2-1}\right)} \\ &= \frac{1}{s} + \frac{\left[2(\zeta^2 - \zeta\sqrt{\zeta^2-1} - 1)\right]^{-1}}{s+\zeta\omega_n - \omega_n\sqrt{\zeta^2-1}} + \frac{\left[2(\zeta^2 + \zeta\sqrt{\zeta^2-1} - 1)\right]^{-1}}{s+\zeta\omega_n + \omega_n\sqrt{\zeta^2-1}} \end{aligned}$$

Taking the Laplace transform inversion, we get the output response in time domain,

$$c(t) = 1 + \frac{1}{2\left(\zeta^2 - \zeta\sqrt{\zeta^2-1} - 1\right)} e^{-\left(\zeta - \sqrt{\zeta^2-1}\right)\omega_n t} + \frac{1}{2\left(\zeta^2 + \zeta\sqrt{\zeta^2-1} - 1\right)} e^{-\left(\zeta + \sqrt{\zeta^2-1}\right)\omega_n t}, \ t \geq 0$$

The response curve is shown in Fig. 6.13(b). Comparing with Fig. 6.13(a), we can find that the overdamped response has a longer response time to the final value.

Example 6.3

A closed-loop TF has the following form:

$$\frac{C}{R} = \frac{9}{s^2 + 4.5s + 9}$$

(1) Determine the undamped natural frequency ω_n, the damping ratio ζ, and the damped natural frequency ω_d.

(2) What is the steady-state output for a unit step input?

Solution (1)

Comparing the closed-loop TF with the generalized form of eq. (6.6),

$$\frac{C}{R} = \frac{\omega_n^2}{s^2 + 2\zeta\omega_n s + \omega_n^2} = \frac{9}{s^2 + 4.5s + 9}$$

We will have

$$\omega_n^2 = 9 \Rightarrow \omega_n = 3$$
$$2\zeta\omega_n = 4.5 \Rightarrow \zeta = 0.75$$

From eq. (6.9), we get

$$\omega_d = \omega_n\sqrt{1 - \zeta^2} = 1.98 \text{ rad/s}$$

Solution (2)

If the system is subjected to a unit step input, the output can be written as

$$C(s) = \frac{9}{s^2 + 4.5s + 9} \cdot R(s) = \frac{9}{s^2 + 4.5s + 9} \cdot \frac{1}{s}$$

By eq. (2.11), applying the final-value theorem yields,

$$\lim_{t\to\infty} c(t) = \lim_{s\to 0} s \cdot C(s) = \lim_{s\to 0} s \cdot \frac{9}{s^2 + 4.5s + 9} \cdot \frac{1}{s} = 1$$

6.2.3 Approximate Analysis of High-Order System

6.2.3.1 Pole and Zero

To analyze a high-order system, we need to recall some technology about the pole and zero. Suppose that the TF $G(s)$ is written in a generalized form, in fact, in the factorized form,

$$G(s) = \frac{K(s+z_1)(s+z_2)\cdots(s+z_m)}{s^k(s+p_1)(s+p_2)\cdots(s+p_n)}$$

The zeros of $G(s)$ are the values that make the numerator zero,

$$K(s+z_1)(s+z_2)\cdots(s+z_m)=0$$

That is,

$$s=-z_1,\ -z_2,\ \cdots,\ -z_m$$

And the poles of $G(s)$ are the values that make the denominator zero,

$$s^k(s+p_1)(s+p_2)\cdots(s+p_n)=0$$

That is,

$$s=-p_1,\ -p_2,\ \cdots,\ -p_n$$

The aforementioned denominator zero equation is also called the characteristic equation. In general, the poles and zeros may be complex, as in the case where the numerator or denominator contains a quadratic or higher-order factor. It is customary to plot the poles and zeros on the complex plane, or s-plane. For the following TF, the quadratic term in the denominator yields a pair of poles at $s = -1 \pm j$,

$$G(s) = \frac{K(s+1)(s+4)}{s(s+2)(s^2+2s+2)}$$

The characteristic equation of $G(s)$ is

$$s(s+2)(s^2+2s+2)=0$$

Thus, we get four poles, including two conjugates, as shown in Fig. 6.14,

$$\begin{array}{ccc} s=0 & \rightarrow & s_1=0 \\ s+2=0 & \rightarrow & s_2=-2 \\ s^2+2s+2=0 & \rightarrow & s_{3,4}=-1\pm j \end{array}$$

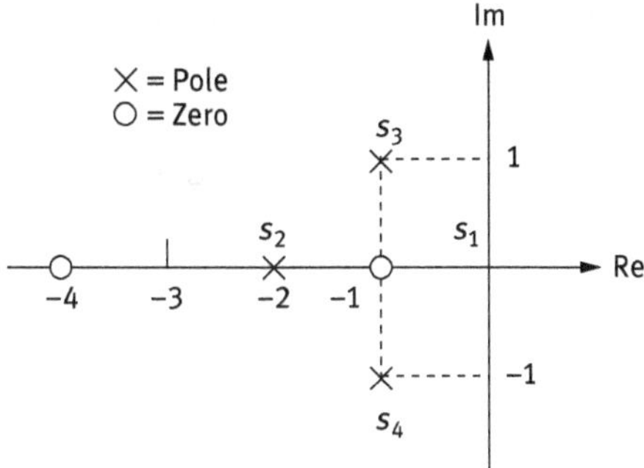

Fig. 6.14: Poles and zeros in s-plane.

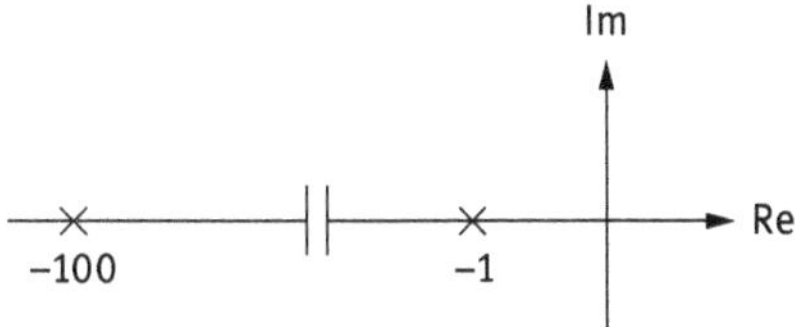

Fig. 6.15: A two-pole system.

6.2.3.2 Dominant Pole

Consider the case where there are two real-axis poles and the TF is given by

$$\frac{C}{R} = \frac{100}{(s+1)(s+100)}$$

In this case, one pole is located close to the imaginary axis at $s = -1$, while the other is located far from the imaginary axis at $s = -100$, as shown in Fig. 6.15.

Assuming that the system is subjected to a unit impulse input $R(s) = 1$, the output may be written as

$$C(s) = \frac{100}{(s+1)(s+100)} \cdot R(s) = \frac{100}{(s+1)(s+100)} \cdot 1 = \frac{A}{s+1} + \frac{B}{s+100}$$

Using the method discussed in Section 2.4.1, we get $A = 100/99$ and $B = -100/99$. Hence, the output can be written as

$$c(t) = 1.01\mathrm{e}^{-t} - 1.01\mathrm{e}^{-100t} = 1.01(\mathrm{e}^{-t} + \mathrm{e}^{-100t})$$

For a given value of t when t tends to infinity, e^{-100t} should be much less than e^{-t}. So

$$c(t) = 1.01\mathrm{e}^{-t} - 1.01\mathrm{e}^{-100t} \approx 1.01\mathrm{e}^{-t}$$

From the above analysis, we can conclude that the poles that are close to the imaginary axis dominate those that are far away, and in many cases the response may be approximated by the response of the poles that are closest to the imaginary axis. Consequently, we call the poles close to the imaginary axis as dominant poles.

We can also get a useful inference of complex poles: the response to any input of a system characterized by two pairs of complex poles at $s_1 = -1 \pm \mathrm{j}$ and $s_2 = -100 \pm \mathrm{j}$ will be dominated by the poles at $s_1 = -1 \pm \mathrm{j}$ because the real part of s_1 is nearer to the imaginary part than the real part of s_2. Hence, s_1 is the dominant pole.

This observation is useful in obtaining an approximate response for higher-order systems by ignoring the contributions from the poles far away from the imaginary axis.

6.2.3.3 Closed-Loop Dominant Poles

To explain the closed-loop dominant poles, we first introduce the term, Bode form TF. The Bode form TF is also called the "+1" form, where the constant term is 1 in the denominator of the closed-loop TF. This is discussed in detail in Chapter 7. Thus, a third-order system in the Bode form TF is represented by

$$\begin{aligned}\frac{C}{R} &= \frac{5}{(\tau s+1)(s^2/\omega_\mathrm{n}^2+2\zeta s\omega_\mathrm{n}+1)} \\ &= \frac{5}{(\tau s+1)\left(s+\zeta\omega_\mathrm{n}-\omega_\mathrm{n}\sqrt{1-\zeta^2}\cdot\mathrm{j}\right)\left(s+\zeta\omega_\mathrm{n}+\omega_\mathrm{n}\sqrt{1-\zeta^2}\cdot\mathrm{j}\right)}\end{aligned}$$

The three roots of this system are as follows, also shown in Fig. 6.16.

$$s_1 = -\frac{1}{\tau},\; s_{2,3} = -\zeta\omega_\mathrm{n} \pm \omega_\mathrm{n}\sqrt{1-\zeta^2}\cdot\mathrm{j}$$

Whether the real pole can be neglected depends on the magnitude of $\zeta\omega_\mathrm{n}$ and $1/\tau$. A common approach is to say that the real pole may be neglected if

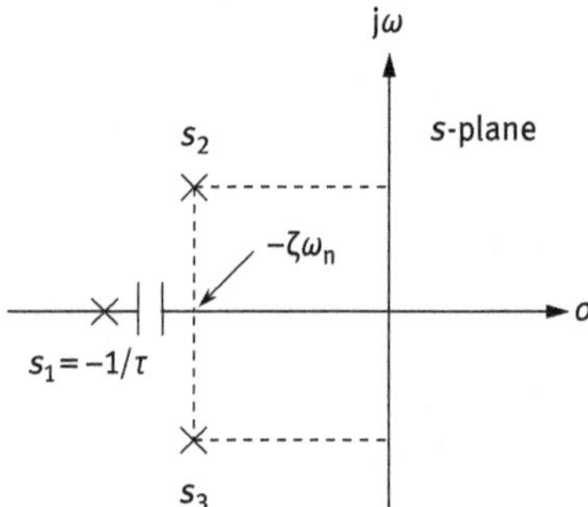

Fig. 6.16: Closed-loop dominant poles of a third-order system.

$$\frac{|1/\tau|}{|\zeta\omega_n|} \geq 5$$

We call those poles, such as $s_{2,3}$, closed-loop dominant poles.

Example 6.4
Reducing the closed-loop TF of the system given below to second-order system,

$$\frac{C}{R} = \frac{600}{(s^2 + 22s + 120)(s^2 + 3s + 4)}$$

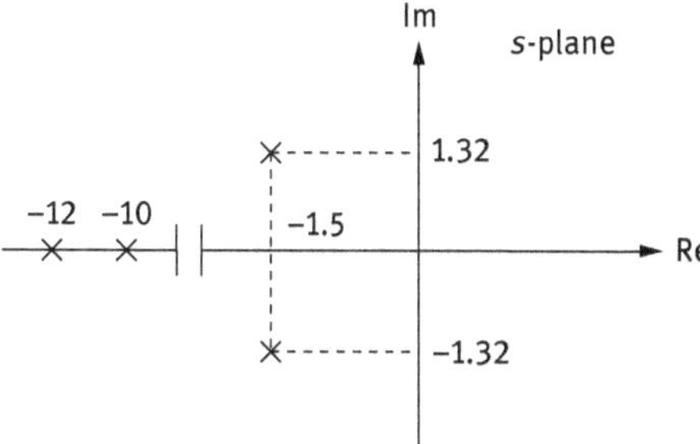

Fig. 6.17: The closed-loop poles.

Solution
Obviously, it is a fourth-order system. First, factorizing the closed-loop TF,

$$\frac{C}{R} = \frac{600}{(s^2 + 22s + 120)(s^2 + 3s + 4)} = \frac{600}{(s + 10)(s + 12)(s + 1.5 + 1.32j)(s + 1.5 - 1.32j)}$$

As can be seen in Fig. 6.17, the four closed-loop poles are

$$s_1 = -10, s_2 = -12, s_3 = -1.5 - 1.32j, s_4 = -1.5 + 1.32j$$

Because

$$\frac{|-12|}{|-1.5|} = 8 \geq 5, \quad \frac{|-10|}{|-1.5|} = 6.7 \geq 5$$

Hence, closed-loop dominant poles are

$$s_{3,4} = -1.5 \pm 1.32j$$

And then this fourth-order system could be reduced to a second-order system,

$$\begin{aligned} \frac{C}{R} &= \frac{600}{\cancel{(s^2 + 22s + 120)}(s^2 + 3s + 4)} \\ \frac{C}{R} &\approx \frac{k}{s^2 + 3s + 4} \end{aligned} \tag{6.11}$$

The gain of system TF should not change before and after reducing, and the gain K before reducing is

$$K=\frac{600}{120\times 4}=1.25$$

Meanwhile, the gain K'after reducing is

$$K'=\frac{k}{4}$$

Finally,

$$K'=K\Rightarrow\frac{k}{4}=1.25\Rightarrow k=5$$

Substituting $k=5$ into eq. (6.11), we can get the final reduced second-order system as

$$\frac{C}{R}\approx\frac{5}{s^2+3s+4}$$

Example 6.5
The closed-loop TF for a system is

$$\frac{Y(s)}{X(s)}=\frac{312000\cdot(s+20.03)}{(s+20)(s+60)(s^2+20s+5200)}$$

Evaluate the approximate unit step input response equation for this system.

Solution
According to the closed-loop TF, the poles are

$$p_{1,2}=-10\pm j71.4, p_3=-60, p_4=-20$$

And the zero of this system is

$$z_1=-20.03$$

Because the pole $p_4=20$ and the zero $z_1=20.03$ are a pair of dipoles, the TF could be simplified to

$$\frac{Y(s)}{X(s)}=\frac{312000\cancel{(s+20.03)}}{\cancel{(s+20)}(s+60)(s^2+20s+5200)}\approx\frac{312000\times 20.03\div 20}{(s+60)(s^2+20s+5200)}$$

About the poles,

$$\begin{matrix}p_{1,2}=-10\pm 71.7j\\ p_3=-60\end{matrix}\Rightarrow\frac{\mathrm{Re}[p_3]}{\mathrm{Re}[p_1]}=\frac{-60}{-10}=6>5$$

Thus, $p_{1,2}$ are the dominant poles and the TF could be simplified again to

$$\frac{Y(s)}{X(s)} = \frac{312000 \times 20.3 \div 20}{(s+60)(s^2+20s+5200)} \approx \frac{312000 \times 20.3 \div 20 \div 60}{s^2+20s+5200} = \frac{k}{s^2+20s+5200}$$

where

$$k = 312000 \times 20.03 \div 20 \div 60 = 5207.8$$

Substituting the K' into the simplified TF, we get

$$\frac{Y(s)}{X(s)} \approx \frac{5207.8}{s^2+20s+5200}$$

Finally, taking the inverse Laplace transform as discussed inSection 2.4.3, we get the approximate unit step input response equation:

$$y(t) \approx 1 - e^{-10t}\sin(71.1t + 1.43)$$

6.3 Performance Specifications in Time Domain

6.3.1 Performance Specifications of First-Order System

For a unit step input signal, the time domain response of first-order system is given by

$$c(t) = 1 - e^{-t/\tau} \tag{6.12}$$

where τ is the time constant. From Fig. 6.18, there are three terminologies related to the first-order system response, the rise time T_r, the settling time T_s,

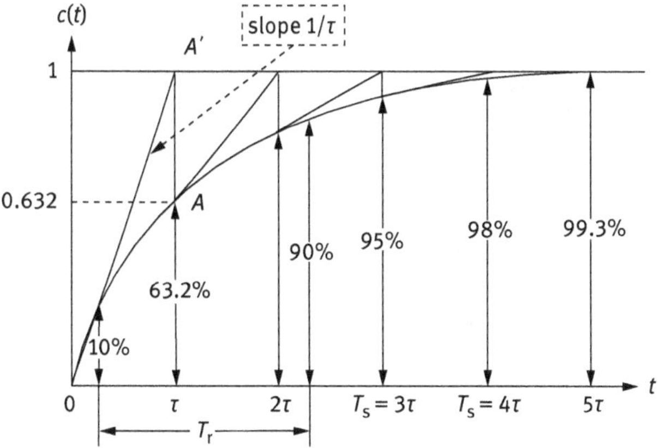

Fig. 6.18: The first-order system response to a unit step.

and the time constant τ. T_s and τ have been mentioned briefly in Section 6.2.1.1. Now we will discuss in details about the three variables.

6.3.1.1 Time constant τ

The time constant is the time when the step response rises to 63.2% of its final value from zero, as shown in Fig. 6.18; in addition, it is the time for $1-e^{-t/\tau}$ to reduce to 36.8% of its initial value,

$$c(\tau) = 1 - e^{-\frac{t}{\tau}}\Big|_{t=\tau} = 1 - e^{-\frac{\tau}{\tau}} = 0.632$$

The derivative of $c(t)$ is $1/\tau$ when $t = 0$, where $1/\tau$ is the slope of $c(t)$ at $t = 0$,

$$\left.\frac{dc(t)}{dt}\right|_{t=0} = \left.\frac{1}{\tau}e^{\frac{-t}{\tau}}\right|_{t=0} = \frac{1}{\tau}$$

The time constant can be considered as a transient response specification for a first-order system, because it is related to the responding speed of the system subjected to a step input.

6.3.1.2 Rise Time T_r

The rise time is the time for the curve to rise from 0.1 to 0.9 of its final value, as shown in Fig. 6.18. Solving eq. (6.2) when $c(t) = 0.9$ and $c(t) = 0.1$,

$$c(t) = 1 - e^{-t/\tau} = 0.9 \Rightarrow t = 2.31\tau$$
$$c(t) = 1 - e^{-t/\tau} = 0.1 \Rightarrow t = 0.11\tau$$

Thus, the rise time T_r should be

$$T_r = 2.31\tau - 0.11\tau = 2.2\tau$$

6.3.1.3 Settling Time T_s

The settling time is the time when the response curve comes to and settles down at ±2% of its final value (other percentages, for example, 5%, can also be used). Let $c(t) =$ 0.98 (or $c(t) = 0.95$) and solving it,

$$c(T_s) = 1 - e^{-T_s/\tau} = 0.98 \Rightarrow T_s = 4\tau(\delta = \pm 2\%)$$

$$c(T_s) = 1 - e^{-T_s/\tau} = 0.95 \Rightarrow T_s = 3\tau(\delta = \pm 5\%)$$

Example 6.6

A unit step first-order response curve is shown in Fig. 6.19. Find the closed-loop TF of this system.

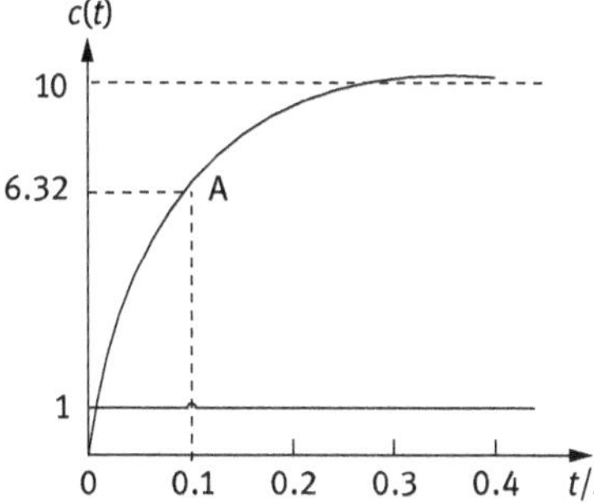

Fig. 6.19: The unit step first-order response curve for a system.

Solution

From Section 6.2.1.4, the general equation of this closed-loop TF is

$$\frac{C(s)}{R(s)} = \frac{k}{\tau s + 1}$$

From the information about point A given in Fig. 6.19, one can get 6.32/10 = 63.2%. So, the time constant $\tau = 0.1$. When time tends to infinity, $c(\infty) = 10$. Thus, according to Section 6.2.1.4, $k \cdot r = 10$. Hence, the gain k of this system should be

$$k = \frac{10}{r} = \frac{10}{1} = 10$$

Finally, the closed-loop TF of this system is

$$G(s) = \frac{k}{\tau s + 1} = \frac{10}{0.1s + 1}$$

6.3.2 Performance Specifications of Second-Order System

For the second-order system, the response swiftness is measured by the rise time, T_r, and the peak time, T_p. For the underdamped system ($0 < \zeta < 1$) with an overshoot, the 0% ~ 100% rise time T_r is a useful index, and not the 10% ~ 90% rise time T_{r1} like in the first-order system. Besides rise time T_r and peak time T_p, there are two other performance specifications of a second-order system, percentage overshoot (PO) and settling time T_s, as shown in Fig. 6.20.

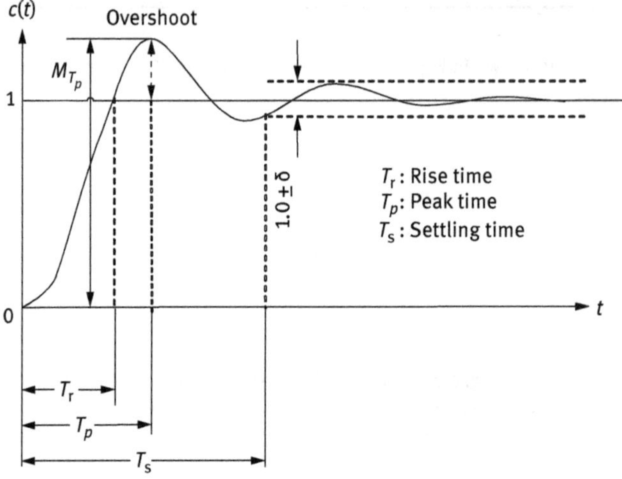

Fig. 6.20: Step response of a second-order system.

6.3.2.1 Rise Time T_r

The rise time T_r is the time when the curve first rises to the final value. For the time response of the second-order system, as mentioned in Section 6.2.2.2,

$$c(t) = 1 - e^{-\zeta\omega_n t}\left(\cos\omega_d t + \frac{\zeta}{\sqrt{1-\zeta^2}}\sin\omega_d t\right)$$

The time when $c(t) = 1$ should be the rise time T_r,

$$c(T_r) = 1 = 1 - e^{-\zeta\omega_n T_r}\left(\cos\omega_d T_r + \frac{\zeta}{\sqrt{1-\zeta^2}}\sin\omega_d T_r\right)$$

$\because$

$$e^{-\zeta\omega_n T_r} \neq 0$$

$\therefore$

$$\cos\omega_d T_r + \frac{\zeta}{\sqrt{1-\zeta^2}}\sin\omega_d T_r = 0$$

Rearranging it,

$$\tan\omega_d T_r = -\frac{\sqrt{1-\zeta^2}}{\zeta}$$

$$\omega_{\mathrm{d}} T_{\mathrm{r}} = \pi - \arctan\frac{\sqrt{1-\zeta^2}}{\zeta}$$

Refering to Fig. 6.11 and let

$$\beta = \arctan\left(\frac{\sqrt{1-\zeta^2}}{\zeta}\right) \tag{6.13}$$

Thus,

$$T_{\mathrm{r}} = \frac{\pi - \beta}{\omega_{\mathrm{d}}} = \frac{\pi - \beta}{\omega_{\mathrm{n}}\sqrt{1-\zeta^2}} \tag{6.14}$$

6.3.2.2 Peak Time T_p

First, we will introduce a terminology M_{T_p}. M_{T_p} is the maximum peak value of the output response curve at the peak time. So, the peak time T_p is the time when the curve first comes to M_{T_p}. For the time response of the second-order system,

$$c(t) = 1 - \mathrm{e}^{-\zeta\omega_{\mathrm{n}}t}\left(\cos\omega_{\mathrm{d}}t + \frac{\zeta}{\sqrt{1-\zeta^2}}\sin\omega_{\mathrm{d}}t\right)$$

The time when one assumes the derivation of $c(t)$ equals to 1, that is, $c'(t) = 0$, is the peak time T_p,

$$\frac{\mathrm{d}c(t)}{\mathrm{d}t}\Big|_{t=T_p} = (1-\zeta^2)\sin\omega_{\mathrm{d}}T_p + \zeta^2\sin\omega_{\mathrm{d}}T_p = 0$$

$$\sin\omega_{\mathrm{d}}T_p = 0$$

That is,

$$\omega_{\mathrm{d}}T_p = n\pi\ ,\ n = 0, 1,2, \ldots,k$$

$$T_p = \frac{\pi}{\omega_{\mathrm{d}}} = \frac{\pi}{\omega_{\mathrm{n}}\sqrt{1-\zeta^2}} \tag{6.15}$$

6.3.2.3 Percentage Overshoot PO

We give the definition of PO as

$$PO = \frac{M_{T_p} - \text{fv}}{\text{fv}} \times 100\% \tag{6.16}$$

where fv is the final value of the response; normally fv is the magnitude of the input. For the second-order system with a unit step input, we have fv = 1. Because M_{T_p} is the maximum peak value of the time response, we have

$$M_{T_p} = 1 - \frac{1}{\sqrt{1-\zeta^2}} e^{-\zeta\omega_n T_p} \cdot \sin\left(\pi + \arctan\frac{\sqrt{1-\zeta^2}}{\zeta}\right) \times 100\%$$

$$\because \quad \beta = \arctan\frac{\sqrt{1-\zeta^2}}{\zeta}$$

$$\begin{aligned}\therefore \quad M_{T_p} &= 1 - \frac{1}{\sqrt{1-\zeta^2}} e^{-\zeta\omega_n T_p} \cdot \sin(\pi - \beta) \\ &= 1 + \frac{1}{\sqrt{1-\zeta^2}} e^{-\zeta\omega_n T_p} \cdot \sin\beta \\ &= 1 + \frac{1}{\sqrt{1-\zeta^2}} e^{-\zeta\omega_n T_p} \cdot \sqrt{1-\zeta^2} \\ &= 1 + e^{-\zeta\omega_n T_p}\end{aligned}$$

From eq. (6.16),

$$PO = \frac{M_{T_p} - \text{fv}}{\text{fv}} \times 100\% = \frac{1 + e^{-\zeta\omega_n T_p} - 1}{1} \times 100\% = e^{-\zeta\omega_n T_p} \times 100\%$$

Together with

$$T_p = \frac{\pi}{\omega_d} = \frac{\pi}{\omega_n\sqrt{1-\zeta^2}}$$

Finally,

$$PO = e^{-\frac{\zeta\pi}{\sqrt{1-\zeta^2}}} \times 100\% \tag{6.17}$$

6.3.2.4 Settling Time T_s

For the second-order system with $0 < \zeta < 1$, the output response curve will be convergent to the input signal. When the difference between the output value and the input value comes to a certain threshold, we consider that the system lies in the steady state. In general, the error to the final value is always chosen as the threshold. We introduce the terminology settling time T_s to analyze this performance for a system. Thus, substituting T_s into the M_{T_p},

$$M_{T_s} = 1 - \frac{1}{\sqrt{1-\zeta^2}} e^{-\zeta\omega_n T_s} \cdot \sin(\pi - \beta) = 1 + e^{-\zeta\omega_n T_s}$$

There are two cases wit regard to T_s. If the response remains within ±2% error of the final value, then

$$M_{T_s} - 1 = 1 + e^{-\zeta\omega_n T_s} - 1 = e^{-\zeta\omega_n T_s} < 0.02$$

That is,

$$\zeta\omega_n T_s \approx 4$$

Hence,

$$T_s = \frac{4}{\zeta\omega_n}(\delta = \pm 2\%) \tag{6.18}$$

If the response remains within ±5% error of the final value, then

$$T_s = \frac{3}{\zeta\omega_n}(\delta = \pm 5\%) \tag{6.19}$$

Example 6.7
Given the closed-loop TF,

$$\Phi(s) = \frac{100}{s^2 + 15s + 100}$$

Find T_p, PO, $T_s(\pm 2\%)$, and T_r .

Solution
Comparing the closed-loop TF with the generalized form,

$$\Phi(s) = \frac{100}{s^2 + 15s + 100} = \frac{{\omega_n}^2}{s^2 + 2\zeta\omega_n s + {\omega_n}^2}$$

One can get

$$\omega_n^2 = 100 \Rightarrow \omega_n = 10 \text{ rad/s}$$
$$2\zeta\omega_n = 15 \Rightarrow \zeta = \frac{15}{2\omega_n} = 0.75$$

From eqs. 6.9 and 6.13,

$$\omega_d = \omega_n \sqrt{1-\zeta^2} = 6.61\,\text{rad/s}$$

$$\beta = \arctan\left(\frac{\sqrt{1-\zeta^2}}{\zeta}\right) \Rightarrow \beta = 0.72$$

Substituting ζ, ω_n, and ω_d into the equations of T_p, PO, T_s, T_r, we get

$$T_p = \frac{\pi}{\omega_d} = \frac{\pi}{\omega_n \sqrt{1-\zeta^2}} = 0.475\text{s}$$

$$\text{PO} = \text{e}^{-\zeta\pi/\sqrt{1-\zeta^2}} \times 100\% = 2.838\%$$

$$T_s = \frac{4}{\zeta\omega_n} = 0.533\text{s}(\delta = \pm 2\%)$$

$$T_r = \frac{\pi - \beta}{\omega_d} = 0.366s$$

Example 6.8

The mechanical vertical translational system is shown in Fig. 6.21(a). When a step input signal with 2*N* amplitude is subjected to this system, the movement rule of the mass *m* is shown in Fig. 6.21(b). Evaluate the values of mass *m*, damping constant *B*, and spring constant *k*.

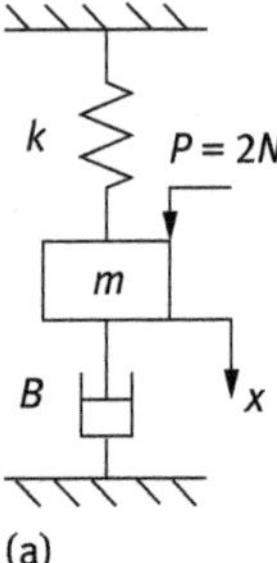

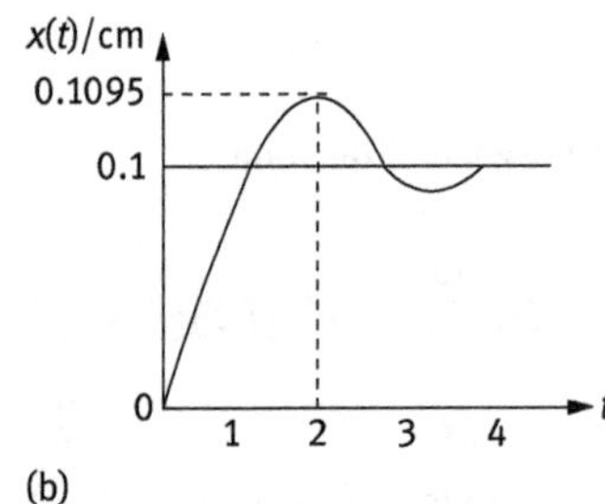

Fig. 6.21: The mechanical vertical translational system and its response curve: (a) the schematic diagram for the example and (b) the response curve for the system in part (a).

Solution

The closed-loop TF of this mechanical vertical system from Fig. 6.21(a) is

$$\frac{X(s)}{P(s)} = \frac{1}{ms^2 + Bs + k}$$

$$\because \quad P(s) = \frac{2}{s}$$

$$\therefore \quad X(s) = \frac{1}{ms^2 + Bs + k} \cdot \frac{2}{s}$$

According to the response curve, the stable or final value of this system is

$$x(\infty) = \lim_{s \to 0} sX(s) = \lim_{s \to 0} s \cdot \frac{1}{ms^2 + Bs + k} \cdot \frac{2}{s} = \frac{2}{k} = 0.001 \text{ m}$$

So,

$$k = \frac{2}{0.001} = 2000 \text{ N/m}$$

From Fig. 6.21(b) and eq. (6.16) for PO,

$$\begin{cases} \text{PO} = \dfrac{0.1095 - 0.1}{0.1}\% = 9.5\% \\ \text{PO} = \text{e}^{-\frac{\zeta\pi}{\sqrt{1-\zeta^2}}} \times 100\% \end{cases}$$

We can get the value of the damping ratio ζ as

$$\zeta = \sqrt{\frac{1}{1 + \left(\frac{\pi}{\ln \text{PO}}\right)^2}} = 0.6$$

Thus,

$$T_p = \frac{\pi}{\omega_n \sqrt{1 - \zeta^2}} = \frac{\pi}{0.8\omega_n} = 2\text{s}$$

$$\omega_n = \frac{3.14}{2 \times 0.8} = 1.96 \text{rad/s}$$

Transforming the closed-loop TF of this mechanical vertical system to make a comparison with the general form of the second-order system,

$$\frac{X(s)}{P(s)} = \frac{1}{ms^2 + Bs + k} = \frac{\frac{1}{m}}{s^2 + \frac{B}{m}s + \frac{k}{m}}$$

So,

$$\omega_n^2 = \frac{k}{m} = \frac{2000}{m}$$

Together with k

$$m = \frac{2000}{\omega_n^2} = \frac{2000}{1.96^2} = 520 \text{ kg}$$

And

$$2\zeta\omega_n = \frac{B}{m}$$

Together with m

$$B = 2\zeta\omega_n m = 2 \times 0.6 \times 1.96 \times 520 = 1220 \text{ N.s/m}$$

6.4 Problems

P6.1. The time response subjected to the unit step input is

$$c(t) = 1 - 2e^{-2t} + e^{-t}$$

Find the closed-loop TF and the unit impulse response.

P6.2. The unit step response curve of the unit negative feedback second-order system is shown in Fig. 6.22. Find its open-loop TF.

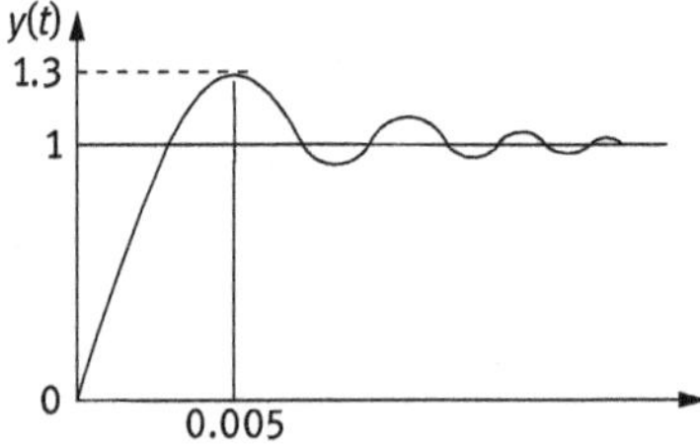

Fig 6.22: The unit step response curve of the unit negative feedback second-order system.

P6.3. For the system shown in Fig. 6.23, if the PO = 16% and $\omega_n = 10$ rad/s when a unit step input is subjected to this system, evaluate the constants a and b.

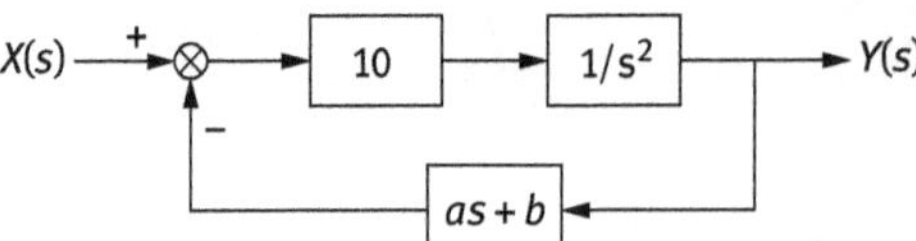

Fig. 6.23: The TF diagram for P6.3.

P6.4. The unit step input response is $c(t) = 3 - 4e^{-t} + e^{-4t}$.

(1) Find the closed-loop TF.

(2) Evaluate the damping ratio ζ and the undamped natural frequency ω_n.

(3) Evaluate the performance specifications T_s, T_p, and PO with $\zeta = 0.707$ and the same value of ω_n in the previous step.

7 Frequency Response Analysis of Control Systems

7.1 Concepts

In the preceding chapters, the time response and performance for systems was in s-plane. A very practical and important alternative approach to analyze and design a system, which is called the frequency response method, provides convenience to investigate the dynamic behavior of control systems and has been widely applied in the domains of analysis and design of control systems.

The frequency response is defined as the steady-state response of a system subjected to a harmonic input signal, namely, a sine-wave signal, whose Euler 's form is

$$i(t) = A_i \sin(\omega t) = \frac{A_i}{2\mathrm{j}}(e^{\mathrm{j}\omega t} - e^{-\mathrm{j}\omega t}) \tag{7.1a}$$

or Laplace transform is

$$I(s) = \frac{A_i}{2\mathrm{j}}\left(\frac{1}{s-\mathrm{j}\omega} - \frac{1}{s+\mathrm{j}\omega}\right) \tag{7.1b}$$

where A_i is the amplitude of sine-wave signal and ω is its frequency. $\mathrm{j} = \sqrt{-1}$. Hence, the system response to a sine-wave input signal can be computed by combining the response to $\mathrm{e}^{p_0 t}$ with $p_0 = \mathrm{j}\omega$ and to $\mathrm{e}^{p_0 t}$ with $p_0 = -\mathrm{j}\omega$.

Let us discuss a linear system represented by transfer function (TF) form:

$$G(s) = \frac{Y(s)}{I(s)} = \frac{K(b_m s^m + b_{m-1}s^{m-1} + \cdots + b_1 s + b_0)}{a_n s^n + a_{n-1}s^{n-1} + \cdots + a_1 s + a_0} \tag{7.2}$$

To simplify the system model, assume that it has different real number poles. Note that evaluating $G(s)$ at s = jω yields a complex number, which can be conveniently represented by its amplitude and phase in polar coordinates,

$$G(s)|_{s=\mathrm{j}\omega} = |G(\mathrm{j}\omega)|\mathrm{e}^{\mathrm{j}\varphi(\omega)} \tag{7.3}$$

Hence, the Laplace transform of the system response can be computed by using a partial fraction expansion to yield the decomposition,

$$Y(s) = G(s) \cdot I(s) = \sum_{i=1}^{n} \frac{K_i}{s-p_i} + \frac{C_1}{s-\mathrm{j}\omega} - \frac{C_2}{s+\mathrm{j}\omega} \tag{7.4}$$

where constant K_i (i = 1,2, ...,n) and C_1, C_2 are as follows:

https://doi.org/10.1515/9783110573275-007

$$K_i = Y(s)(s-p_i)|_{s=p_i}$$

$$C_1 = Y(s)(s-\mathrm{j}\omega)|_{s=\mathrm{j}\omega} = \left.\frac{A_i G(s)}{2\mathrm{j}}\right|_{s=\mathrm{j}\omega} = \frac{A_i|G(\mathrm{j}\omega)|}{2\mathrm{j}}\mathrm{e}^{\mathrm{j}\varphi(\omega)} = \frac{A_i A(\omega)}{2\mathrm{j}}\mathrm{e}^{\mathrm{j}\varphi(\omega)}$$

$$C_2 = Y(s)(s+\mathrm{j}\omega)|_{s=-\mathrm{j}\omega} = \left.\frac{A_i G(s)}{2\mathrm{j}}\right|_{s=-\mathrm{j}\omega} = \frac{A_i|G(-\mathrm{j}\omega)|}{2\mathrm{j}}\mathrm{e}^{-\mathrm{j}\varphi(\omega)}$$

$$= \frac{A_i|G(\mathrm{j}\omega)|}{2\mathrm{j}}\mathrm{e}^{-\mathrm{j}\varphi(\omega)} = \frac{A_i A(\omega)}{2\mathrm{j}}\mathrm{e}^{-\mathrm{j}\varphi(\omega)}$$

The system response to a sine-wave input can be computed as follows:

$$y(t) = \sum_{i=1}^{n} K_i \mathrm{e}^{p_i t} + \frac{A_i|G(\mathrm{j}\omega)|}{2\mathrm{j}}\left[\mathrm{e}^{\mathrm{j}[\omega t+\varphi(\omega)]} - \mathrm{e}^{-\mathrm{j}[\omega t+\varphi(\omega)]}\right]$$

$$= \sum_{i=1}^{n} K_i \mathrm{e}^{p_i t} + A_i|G(\mathrm{j}\omega)|\sin[\omega t+\varphi(\omega)]$$

when $p_i < 0$ $(i = 1,2,\dots,n)$, the term of $\sum_{i=1}^{n} K_i \mathrm{e}^{p_i t}$ tends to be zero, the system steady-state response becomes

$$y(t) = A_i|G(\mathrm{j}\omega)|\sin\left[\omega t+\varphi_y(\omega)\right] \tag{7.5}$$

Thus, we conclude that a sine-wave input signal forces a sine wave at the output signal with the same frequency for a linear system. Moreover, the amplitude of the output sine wave is modified by a factor equal to the gain of $A_i\,|G(\mathrm{j}\omega)\,|$ and the phase is shifted by a quantity equal to the phase of $G(\mathrm{j}\omega)$, shown in Fig. 7.1.

When the TF $G(s)$ is evaluated at $s = \mathrm{j}\omega$, a complex number $G(\mathrm{j}\omega)$, called the frequency characteristic function, is defined, as shown in Fig. 7.2.

Its expression is as follows:

$$G(\mathrm{j}\omega) = \mathrm{Re}(\omega) + \mathrm{j}\mathrm{Im}(\omega) = |G(\mathrm{j}\omega)|\mathrm{e}^{\mathrm{j}\varphi(\omega)} = A(\omega)\cdot \mathrm{e}^{\mathrm{j}\varphi(\omega)} \tag{7.6}$$

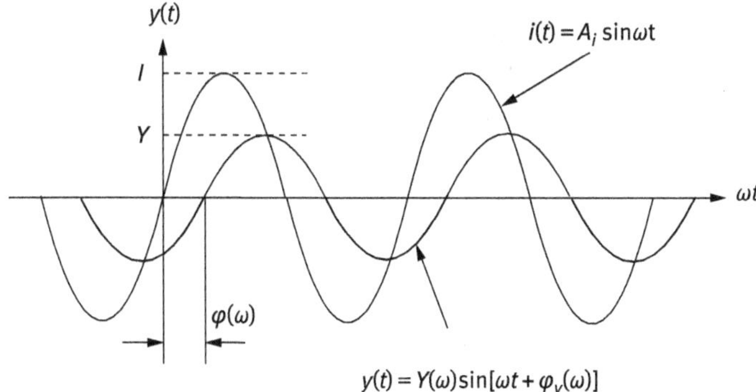

Fig. 7.1: The system response subjected to a sine-wave input signal.

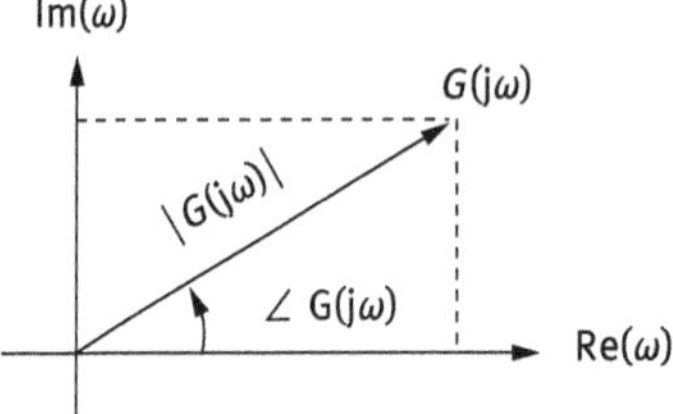

Fig. 7.2: The graphical representation of frequency characteristic.

Where $\mathrm{Re}(\omega)$ is the real part of $G(\mathrm{j}\omega)$, which is called the real frequency characteristic function, and $\mathrm{Im}(\omega)$ is the imaginary part of $G(\mathrm{j}\omega)$, which is called the imaginary frequency characteristic function. The amplitude $|G(\mathrm{j}\omega)|$, which is identical to $A(\omega)$, and phase $\varphi(\omega)$ are, respectively, called the amplitude frequency characteristic function and the phase frequency characteristic function and are represented by

$$|G(\mathrm{j}\omega)| = \sqrt{\mathrm{Re}(\omega)^2 + \mathrm{Im}(\omega)^2} \tag{7.7}$$

$$\varphi(\omega) = \arctan\frac{\mathrm{Im}(\omega)}{\mathrm{Re}(\omega)} \tag{7.8}$$

From eq. (7.5), the amplitude and phase of the frequency response are, respectively, related to $|G(\mathrm{j}\omega)|$ and $\varphi(\omega)$ of the frequency characteristic function for the linear system. Comparing eqs. (7.5) and (7.1a), one can write these relationships as

$$|G(\mathrm{j}\omega)| = \frac{|y(t)|}{|i(t)|} \tag{7.9}$$

and

$$\varphi(\omega) = \varphi_y(\omega) - \varphi_i(\omega) \tag{7.10}$$

where $|y(t)|$ is the amplitude of the frequency response output, $|i(t)|$ is the amplitude of sine-wave input signal, $\varphi_y(\omega)$ is the phase of the frequency response output, and $\varphi_i(\omega)$ is the phase of sine-wave input signal.

The amplitude frequency characteristic indicates whether the amplitude of the system response output is amplified or reduced. The phase frequency characteristic shows whether the system response output signal exceeds ($\varphi(\omega) > 0$) or postpones ($\varphi(\omega) < 0$) to the sine-wave input signal.

Example 7.1

Determine the frequency characteristic functions when given the TF is

$$G(s) = \frac{K}{1 + Ts}$$

Solution

Substitute $s = \mathrm{j}\,\omega$ into the equation and split it into real and imaginary parts, so that

$$
\begin{aligned}
G(j\omega) &= \frac{K}{1+Tj\omega} \\
&= \frac{K}{(1+Tj\omega)(1-Tj\omega)}(1-j\omega T) \\
&= \frac{K}{1+(T\omega)^2}(1-j\omega T) \\
&= \frac{K}{1+(T\omega)^2} - \frac{K\omega T}{1+(T\omega)^2}j
\end{aligned}
$$

where

$$\mathrm{Re}(\omega) = \frac{K}{1+(T\omega)^2}$$

$$\mathrm{Im}(\omega) = \frac{-K\omega T}{1+(T\omega)^2}$$

furthermore

$$|G(j\omega)| = \frac{K}{\sqrt{1+(T\omega)^2}}$$

$$\varphi(\omega) = \arctan(-\omega T)$$

Example 7.2

Consider the frequency characteristic function, including the amplitude and phase frequency characteristic functions, and according to the TF it is given by

$$G(s) = \frac{s+1}{s^2+5s+6}$$

Solution

Its zeros and poles are $z = -1$, and $p_1 = -2$, $p_2 = -3$, so that

$$G(s) = \frac{s+1}{(s+2)(s+3)}$$

Let $s = j\,\omega$, and so we obtain

$$G(j\omega) = \frac{j\omega+1}{(j\omega+2)(j\omega+3)}$$

According to the computation rules of complex number, the amplitude and the phase frequency characteristic functions are

$$|G(j\omega)| = \frac{|j\omega + 1|}{|j\omega + 2| \cdot |j\omega + 3|} = \frac{\sqrt{1+\omega^2}}{\sqrt{2^2+\omega^2} \cdot \sqrt{3^2+\omega^2}}$$

$$\varphi(\omega) = \arctan \omega - \arctan\frac{\omega}{2} - \arctan\frac{\omega}{3}$$

Example 7.3

Given input signal $i(t) = 2\sin(t + 45°)$, consider the TF given by

$$G(s) = \frac{Y(s)}{I(s)} = \frac{s+1}{(s+2)(s+4)}$$

Determine the frequency response output.

Solution

Let $s = j\,\omega$, then

$$|G(j\omega)| = \frac{|1 + j\omega|}{|j\omega + 2| \cdot |j\omega + 4|} = \frac{\sqrt{1+\omega^2}}{\sqrt{2^2+\omega^2} \cdot \sqrt{4^2+\omega^2}}$$

$$\varphi(\omega) = \arctan \omega - \arctan\frac{\omega}{2} - \arctan\frac{\omega}{4}$$

Because the input signal is $i(t) = 2\sin(t + 45°)$, we can get $\omega = 1$. When $\omega = 1$, the amplitude frequency characteristic function $|G(j\omega)|$ is as follows:

$$|G(j\omega)|\big|_{\omega=1} = \frac{\sqrt{2}}{\sqrt{5} \cdot \sqrt{17}} = 0.1534$$

and the phase frequency characteristic function $\varphi(\omega)$ is

$$\varphi(\omega)|_{\omega=1} = \arctan 1 - \arctan\frac{1}{2} - \arctan\frac{1}{4} = 4.4°$$

According to eqs. (7.9) and (7.10), the gain and phase of the frequency response output of the system are as follows:

$$|y(t)| = |G(j\omega)| \cdot |i(t)| = 0.1534 \times 2 = 0.3068$$

$$\varphi_y(\omega) = \varphi(\omega) + \varphi_i(\omega) = 4.4° + 45° = 49.4°$$

Therefore, the frequency response output signal is

$$y(t) = 0.3068 \sin(t + 49.4°)$$

7.2 Graphical Description: Nyquist Diagram and Bode Diagram

The frequency response is a very useful tool for all aspects of analysis, synthesis, and design of controllers. Because of its importance, special diagrams are introduced to plot frequency response. They are typically depicted graphically in the form of a polar diagram (commonly called the Nyquist diagram) or in both gain and phase form (commonly called the Bode diagram). More will be discussed about the Nyquist diagram and Bode diagram in this section.

7.2.1 Simple Rules for Plotting a Nyquist Diagram

Take *X*-axis and *Y*-axis as the real part and the imaginary part of the frequency characteristic function, respectively, in the $X-Y$ reference frame. When frequency ω changes in $(-\infty, +\infty)$, the graph of $G(\mathrm{j}\omega)$ is called the Nyquist diagram or the polar coordinate diagram.

Fig. 7.3 shows the Nyquist diagram for a vector $G(\mathrm{j}\omega)$ and its coordinates of point $G(\mathrm{Re}(\omega), \mathrm{Im}(\omega))$ for a given frequency ω. Therefore, the diagram can be obtained by tracing different frequency ω the Nyquist.

Main rules for plotting a Nyquist diagram are given as follows:

(1) Put $s = \mathrm{j}\,\omega$ into TF of the system to obtain the frequency characteristic $G(\mathrm{j}\omega)$.

(2) Write down the amplitude frequency characteristic function $|G(\mathrm{j}\omega)|$, phase frequency characteristic function $\angle G(\mathrm{j}\omega)$, real frequency characteristic function $\mathrm{Re}(\omega)$, and imaginary frequency characteristic function $\mathrm{Im}(\omega)$.

(3) Let $\omega = 0$ and get $|G(\mathrm{j}\omega)|$, $\angle G(\mathrm{j}\omega)$, $\mathrm{Re}(\omega)$, and $\mathrm{Im}(\omega)$ when $\omega = 0$.

(4) Let $\mathrm{Re}(\omega) = 0$ and obtain the value of ω. Substituting the value of ω into $\mathrm{Im}(\omega)$ we get the intersection of the end of the vector $G(\mathrm{j}\omega)$ and imaginary axis. Similarly, let $\mathrm{Im}(\omega) = 0$ and we get the value of ω. Substituting the value of ω into $\mathrm{Re}(\omega)$ we get the intersection of the end of the vector $G(\mathrm{j}\omega)$ and real axis.

(5) For the second-order oscillation system, get $|G(\mathrm{j}\omega_n)|$, $\angle G(\mathrm{j}\omega_n)$, $\mathrm{Re}(\omega_n)$, and $\mathrm{Im}(\omega_n)$ when $\omega = \omega_n$.

(6) Let $\omega = \infty$ and get $|G(\mathrm{j}\omega_\infty)|$, $\angle G(\mathrm{j}\omega_\infty)$, $\mathrm{Re}(\omega_\infty)$, and $\mathrm{Im}(\omega_\infty)$.

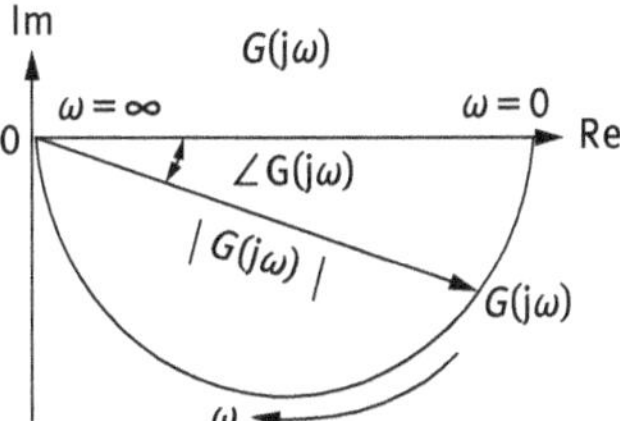

Fig. 7.3: The Nyquist diagram.

(7) Besides the aforementioned special values of ω, we can select some different values of ω according to $0 < \omega < \infty$ to evaluate $|G(\mathrm{j}\omega)|$, $\angle G(\mathrm{j}\omega)$, $\mathrm{Re}(\omega)$, and $\mathrm{Im}(\omega)$ to get a precise curve.

(8) In the s-plane, we draw the real axis, original point, and imaginary axis. Then, connecting the above points according to the value of ω from 0 to infinity we get a curve. Finally, we obtain the direction of ω from 0 to infinity.

7.2.2 The Nyquist Diagrams for Typical Links

7.2.2.1 Proportion Link

The TF of the proportion link is

$$G(s) = K$$

Thus, the frequency characteristic function is

$$G(\mathrm{j}\omega) = K \tag{7.11}$$

Thus, the real frequency characteristic of the proportion link is $\mathrm{Re}(\omega) = K$, and the imaginary frequency characteristic is $\mathrm{Im}(\omega) = 0$. The amplitude frequency characteristic is $|G(\mathrm{j}\omega)| = K$, and the phase frequency characteristic $\angle G(\mathrm{j}\omega) = 0°$. When ω changes from 0 to ∞, $G(\mathrm{j}\omega)$ is always a point $(K, \mathrm{j}0)$ on the real axis of the s-plane, as shown in Fig. 7.4.

7.2.2.2 Integral Link

The TF of the integral link is

$$G(s) = \frac{1}{s}$$

Hence, the frequency characteristic function is

$$G(\mathrm{j}\omega) = \frac{1}{\mathrm{j}\omega} = -\mathrm{j}\frac{1}{\omega} \tag{7.12}$$

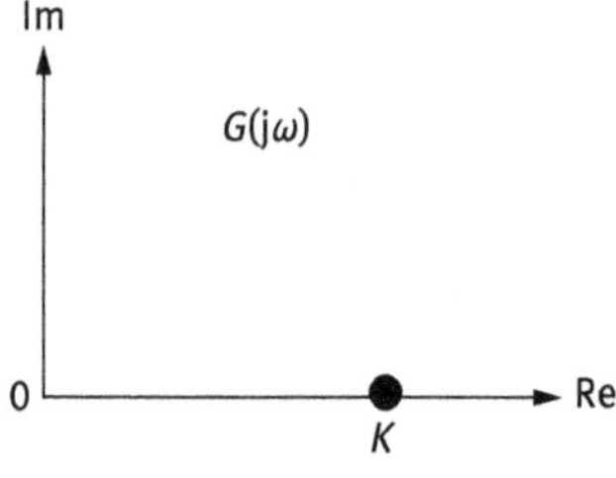

Fig. 7.4: The Nyquist diagram for the proportion link.

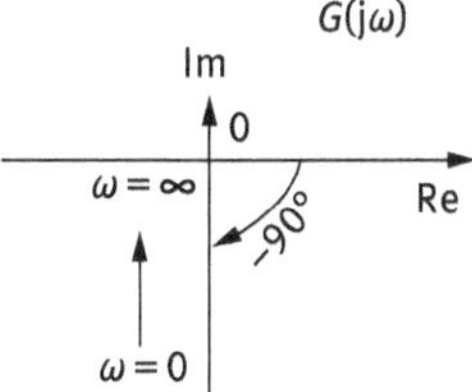

Fig. 7.5: The Nyquist diagram for the integral link.

Thus, the real frequency characteristic of the integral link is $\text{Re}(\omega) = 0$, and the imaginary frequency characteristic is $\text{Im}(\omega) = -1/\omega$. The amplitude frequency characteristic is $|G(\text{j}\omega)| = 1/\omega$, and the phase frequency characteristic $\angle G(\text{j}\omega) = -90°$. When ω changes from 0 to ∞, the amplitude of $G(\text{j}\omega)$ changes from ∞ to 0 and the phase of $G(\text{j}\omega)$ is at $-90°$. In the Nyquist diagram of the integral link, $G(\text{j}\omega)$ is the bottom half of the imaginary axis, whose direction is from $-\infty$ to the original point, as shown in Fig. 7.5.

7.2.2.3 Ideal Differential Link

The TF of the ideal differential link is

$$G(s) = s$$

Thus, the frequency characteristic function is

$$G(\text{j}\omega) = \text{j}\omega \tag{7.13}$$

Thus, the real frequency characteristic of the ideal differential link is $\text{Re}(\omega) = 0$, and the imaginary frequency characteristic is $\text{Im}(\omega) = \omega$. The amplitude frequency characteristic is $|G(\text{j}\omega)| = \omega$, and the phase frequency characteristic $\angle G(\text{j}\omega) = 90°$. When ω changes from 0 to ∞, the amplitude of $G(\text{j}\omega)$ changes from 0 to ∞ and the phase of $G(\text{j}\omega)$ is at 90°. The Nyquist diagram of the ideal differential link, $G(\text{j}\omega)$ is the upper half of the imaginary axis, whose direction is from the original point to ∞, as shown in Fig. 7.6.

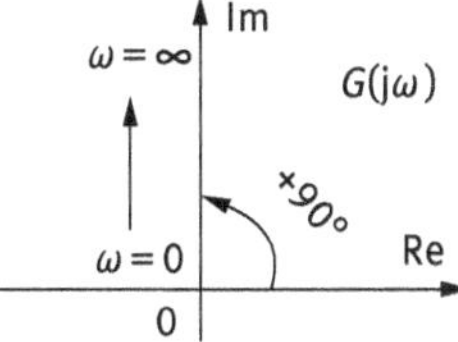

Fig. 7.6: The Nyquist diagram for the ideal differential link.

7.2.2.4 Inertial Link

The TF of the inertial link is

$$G(s) = \frac{1}{Ts + 1}$$

Hence, the frequency characteristic function is

$$G(\mathrm{j}\omega)=\frac{1}{1+\mathrm{j}\omega T}=\frac{1}{1+\omega^2T^2}-\frac{T\omega}{1+\omega^2T^2}\cdot\mathrm{j} \tag{7.14}$$

Thus, the real and the imaginary frequency characteristics are, respectively,

$$\mathrm{Re}(\omega)=\frac{1}{1+\omega^2T^2}$$

$$\mathrm{Im}(\omega)=\frac{-T\omega}{1+\omega^2T^2}$$

The amplitude and the phase frequency characteristics are, respectively,

$$|G(\mathrm{j}\omega)|=\frac{1}{\sqrt{1+T^2\omega^2}}$$

$$\angle G(\mathrm{j}\omega)=-\arctan T\omega$$

When $\omega=0$, $|G(\mathrm{j}\omega)|=1$, $\angle G(\mathrm{j}\omega)=0°$, $\mathrm{Re}(\omega)=1$, $\mathrm{Im}(\omega)=0$.
When $\omega=\frac{1}{T}$, $|G(\mathrm{j}\omega)|=0.707$, $\angle G(\mathrm{j}\omega)=-45°$, $\mathrm{Re}(\omega)=\frac{1}{2}$, $\mathrm{Im}(\omega)=-\frac{1}{2}$.
When $\omega=\infty$, $|G(\mathrm{j}\omega)|=0$, $\angle G(\mathrm{j}\omega)=-90°$, $\mathrm{Re}(\omega)=0$, $\mathrm{Im}(\omega)=0$.

Using new variable U and V to represent the real frequency characteristic and imaginary frequency characteristics, respectively,

$$U=\frac{1}{1+\omega^2T^2},\quad V=\frac{-T\omega}{1+\omega^2T^2}$$

one can get

$$\left(U-\frac{1}{2}\right)^2+V^2=\left(\frac{1}{2}\right)^2$$

The aforementioned equation represents a circle. $\angle G(\mathrm{j}\omega)$ and $\mathrm{Im}(\omega)$ are always negative when ω changes from 0 to ∞, and so the Nyquist diagram of the inertial link should be the bottom half circle, as shown in Fig. 7.7.

7.2.2.5 First-Order Differential Link

The TF of the first-order differential link is

$$G(s)=Ts+1$$

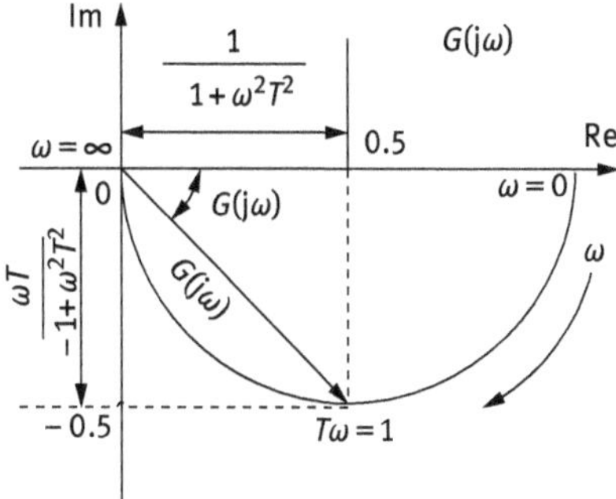

Fig. 7.7: The Nyquist diagram for the inertial link.

Thus, the frequency characteristic function is

$$G(\mathrm{j}\omega) = 1 + \mathrm{j}\omega T \tag{7.15}$$

So, the real frequency characteristic of the first-order differential link is

$$\mathrm{Re}(\omega) = 1$$

And the imaginary frequency characteristic is

$$\mathrm{Im}(\omega) = T\omega$$

The amplitude frequency characteristic is

$$|G(\mathrm{j}\omega)| = \sqrt{1 + T^2\omega^2}$$

The phase frequency characteristic is

$$\angle G(\mathrm{j}\omega) = \arctan T\omega$$

When $\omega = 0$, Re = 1, Im = 0, $|G(\mathrm{j}\omega)| = 1$, $\angle G\,(\mathrm{j}\omega) = 0°$.
When $\omega = 1/T$, Re = 1, Im = 1, $|G(\mathrm{j}\omega)| = 2^{1/2}$, $\angle G(\mathrm{j}\omega) = 45°$.
When $\omega = \infty$, Re = 1, Im = ∞, $|G(\mathrm{j}\omega)| = \infty$, $\angle G(\mathrm{j}\omega) = 90°$.

So from the aforementioned analysis, we can conclude that when ω changes from 0 to ∞, the amplitude of $G(\mathrm{j}\omega)$ changes from 1 to ∞ and the phase of $G(\mathrm{j}\omega)$ changes from 0° to 90°. The Re(ω) is always 1. Therefore, the Nyquist diagram of the first-order differential link is a line, which is perpendicular to real axis with the starting point (1, j0), as shown in Fig. 7.8.

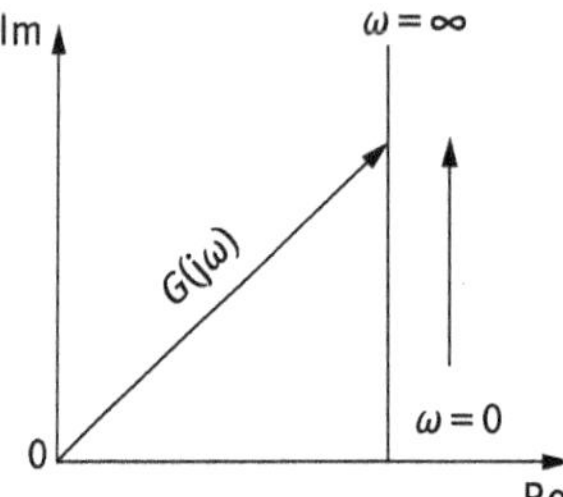

Fig. 7.8: The Nyquist diagram for the first-order differential link.

7.2.2.6 Second-Order Oscillation Link

The TF of the second-order oscillation link is

$$G(s) = \frac{\omega_n^2}{s^2 + 2\zeta\omega_n s + \omega_n^2}, \qquad 0 < \zeta < 1$$

Transferring the denominator into the form of "+1", and let $T = 1/\ \omega_n$,

$$G(s) = \frac{1}{T^2 s^2 + 2\zeta T s + 1}$$

Let $s = \mathrm{j}\omega$; the frequency characteristic function is

$$\begin{aligned} G(\mathrm{j}\omega) &= \frac{1}{T^2(\mathrm{j}\omega)^2 + 2\zeta T(\mathrm{j}\omega) + 1} \\ &= \frac{1 - T^2\omega^2}{(1 - T^2\omega^2)^2 + (2\zeta T\omega)^2} - \mathrm{j}\frac{2\zeta T\omega}{(1 - T^2\omega^2)^2 + (2\zeta T\omega)^2} \end{aligned} \tag{7.16}$$

Thus, the real frequency characteristic of the second-order oscillation link is

$$\mathrm{Re}(\omega) = \frac{1 - T^2\omega^2}{(1 - T^2\omega^2)^2 + (2\zeta T\omega)^2}$$

And the imaginary frequency characteristic is

$$\mathrm{Im}(\omega) = -\frac{2\zeta T\omega}{(1 - T^2\omega^2)^2 + (2\zeta T\omega)^2}$$

The amplitude frequency characteristic is

$$|G(\mathrm{j}\omega)| = \frac{1}{\sqrt{(1 - T^2\omega^2)^2 + (2\zeta T\omega)^2}}$$

The phase frequency characteristic is

$$\angle G(j\omega) = -\arctan\frac{2\zeta T\omega}{1-T^2\omega^2}$$

On the basis of the aforementioned analysis, one can find

when $\omega = 0$, $\mathrm{Re} = 1, \mathrm{Im} = 0$, $|G(j\omega)| = 1$, $\angle G(j\omega) = 0°$,
when $\omega = \frac{1}{T}$, $\mathrm{Re} = 0, \mathrm{Im} = -\frac{1}{2\zeta}$, $|G(j\omega)| = \frac{1}{2\zeta}$, $\angle G(j\omega) = -90°$,
when $\omega = \infty$, $\mathrm{Re} = 0, \mathrm{Im} = 0$, $|G(j\omega)| = 0$, $\angle G(j\omega) = -180°$.

The Nyquist diagram for the second-order oscillation link starts at the point (1, j0) and ends at the point (0, j0), as shown in Fig. 7.9. The frequency of the intersection of the curve and the imaginary axis is ω_n, and the amplitude of the intersection is $1/(2\,\zeta)$. And when ω changes from 0 to ∞, $|G(j\omega)\,|$ changes from 1 to 0 and $\angle G(j\omega)$ changes from $0° \sim 180°$.

If $|G(j\omega)\,|$ has a peak value (extremum), which is called resonance peak value M_r, at resonance frequency ω_r, we will have

$$\left.\frac{\partial|G(j\omega)|}{\partial\omega}\right|_{\omega=\omega_r} = 0$$

Substituting the amplitude frequency characteristic equation into the above equation, we have

$$\omega_r = \frac{1}{T}\sqrt{1-2\zeta^2} = \omega_n\sqrt{1-2\zeta^2} \tag{7.17}$$

If ω_r is meaningful or existent, it must be $1-2\,\zeta^2 \geq 0$; in other words, it must be $0 < \zeta \leq 0.707$. Only when $0 < \zeta \leq 0.707$, there would be a resonance peak value M_r,

$$M_r = |G(j\omega_r)| = \frac{1}{2\zeta\sqrt{1-\zeta^2}} \tag{7.18}$$

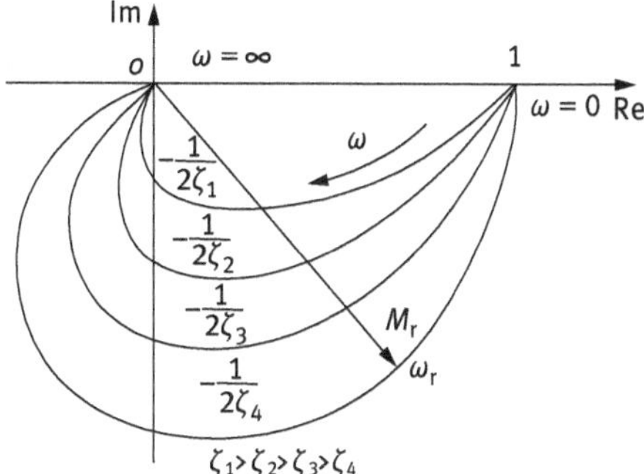

Fig. 7.9: The Nyquist diagram for the second-order oscillation link.

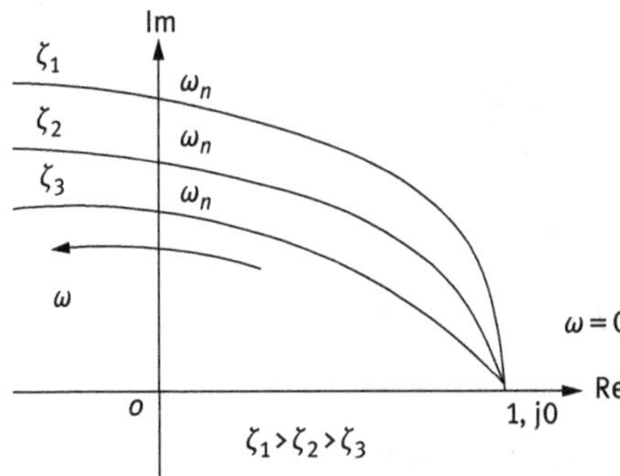

Fig. 7.10: The Nyquist diagram for the second-order differential link.

Consequently, the amplitude frequency characteristic is

$$\angle G(\mathrm{j}\omega_r) = -\arctan\frac{\sqrt{1-2\zeta^2}}{\zeta} \tag{7.19}$$

7.2.2.7 Second-Order Differential Link

The TF of the second-order differential link is

$$G(s) = T^2 s^2 + 2\zeta T s + 1,\ T = \frac{1}{\omega_n}$$

Let $s = \mathrm{j}\,\omega$, the frequency characteristic function is

$$G(\mathrm{j}\omega) = T^2(\mathrm{j}\omega)^2 + 2\zeta T(\mathrm{j}\omega) + 1 \tag{7.20}$$

The Nyquist diagram for the second-order differential link is shown in Fig. 7.10.

Example 7.4

Plot the Nyquist diagram for the system with the TF,

$$G(s) = \frac{100}{0.01s^2 + 0.1s + 1}$$

Solution

According the given function and the equation of the second-order oscillation link, one can get

$$T = 0.1,\ \zeta = 0.5,\ \omega_n = 10$$

Let $s = \mathrm{j}\omega$, the frequency characteristic function is

$$G(\mathrm{j}\omega) = \frac{100}{1 + 0.1\mathrm{j}\omega - 0.01\omega^2} = \frac{100 - \omega^2}{(1 - 0.01\omega^2)^2 + 0.01\omega^2} - \mathrm{j}\frac{10\omega}{(1 - 0.01\omega^2)^2 + 0.01\omega^2}$$

Thus,

$$\mathrm{Re}(\omega) = \frac{100 - \omega^2}{(1 - 0.01\omega^2)^2 + 0.01\omega^2}$$

$$\mathrm{Im}(\omega) = -\frac{10\omega}{(1 - 0.01\omega^2)^2 + 0.01\omega^2}$$

$$|G(\mathrm{j}\omega)| = \frac{100}{\sqrt{(1 - 0.01\omega^2)^2 + 0.01\omega^2}}$$

$$\angle G(\mathrm{j}\omega) = -\arctan\frac{10\omega}{100 - \omega^2}$$

when $\omega = 0$, we have

$$\begin{cases} |G(\mathrm{j}\omega)| = 100 \\ \angle G(\mathrm{j}\omega) = 0 \\ \mathrm{Re}(\omega) = 100 \\ \mathrm{Im}(\omega) = 0 \end{cases}$$

when $\omega = \omega_n = 10$, we have

$$\begin{cases} |G(\mathrm{j}\omega)| = 100 \\ \angle G(\mathrm{j}\omega) = -90^\circ \\ \mathrm{Re}(\omega) = 0 \\ \mathrm{Im}(\omega) = -100 \end{cases}$$

when $\omega = \infty$, we have

$$\begin{cases} |G(\mathrm{j}\omega)| = 0 \\ \angle G(\mathrm{j}\omega) = -180^\circ \\ \mathrm{Re}(\omega) = 0 \\ \mathrm{Im}(\omega) = 0 \end{cases}$$

Because $\zeta = 0.5 < 0.707$, the resonance peak value M_r should exist. Substituting ζ into eq. (7.17), we have

$$\omega = \omega_r = \omega_n\sqrt{1 - 2\zeta^2} = 10\sqrt{1 - 2 \times 0.5^2} = 5\sqrt{2}$$

And substituting ω_r into eqs. (7.18) and (7.19), we have

$$\begin{cases} M_r = |G(j\omega_r)| = \frac{k}{2\zeta\sqrt{1-\zeta^2}} = \frac{100}{2\zeta\sqrt{1-\zeta^2}} = 115 \\ \angle G(j\omega_r) = -\arctan\frac{\sqrt{1-2\zeta^2}}{\zeta} = -54.7^\circ \end{cases}$$

The Nyquist diagram for this example is shown in Fig. 7.11.

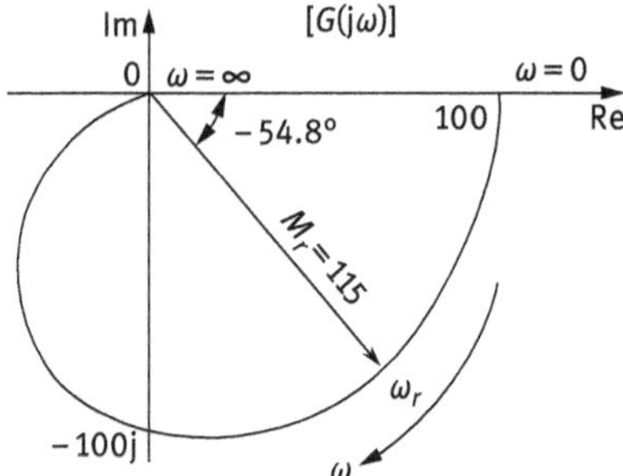

Fig. 7.11: The Nyquist diagram for Example 7.4.

7.2.3 Simple Rules for Plotting Bode Diagrams

The Bode diagram is also called the logarithm frequency characteristic diagram, and consists of two diagrams: the logarithm gain/amplitude characteristic diagram and the logarithm phase characteristic diagram. They show characteristics of gain and change of phase of $G(\mathrm{j}\omega)$ along with movement of frequency ω, respectively, and logarithm scales are usually used for the frequency axis ω and the gain of $G(\mathrm{j}\omega)$.

Taking logarithm transfer for frequency characteristic of any link, we get

$$\ln G(\mathrm{j}\omega) = \ln[|G(\mathrm{j}\omega)| \cdot \mathrm{e}^{\mathrm{j}\varphi(\omega)}] = \ln|G(\mathrm{j}\omega)| + \mathrm{j}\varphi(\omega) \tag{7.21}$$

On the basis of eq. (7.21), we have following explanations with details:

(1) The real part is also called the logarithm amplitude frequency characteristic with the form of $\ln|G(\mathrm{j}\omega)|$.
(2) The imaginary part is also called the logarithm phase frequency characteristic with the form of $\varphi(\omega)$. The imaginary part depicts the fact that $\angle G(\mathrm{j}\omega)$ varies along with movement of frequency ω.
(3) In practical applications, lg is always used in logarithm amplitude frequency characteristic diagram instead of ln,

$$L(\omega) = 20\lg|G(\mathrm{j}\omega)| \tag{7.22}$$

(4) Logarithm scales are usually used for the frequency axis ω and the amplitude of $G(\mathrm{j}\omega)$.
(5) The unit of $L(\omega)$ is decibel (dB). For example, if

$$20\lg\frac{N_2}{N_1} = 1\ \mathrm{dB}$$

That is, N_2 is more than N_1. Furthermore, the difference of N_2 and N_1 is 1 dB.

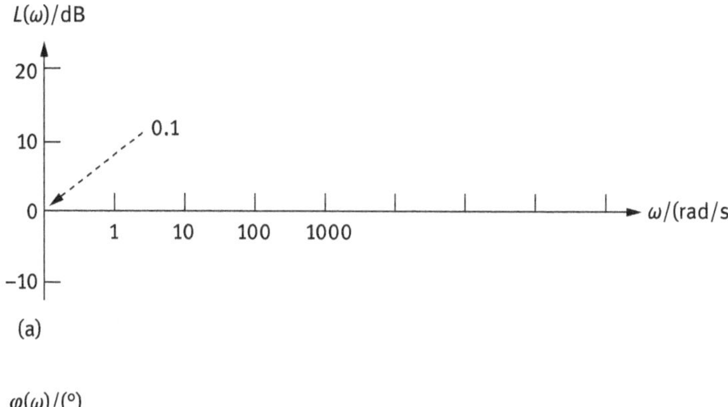

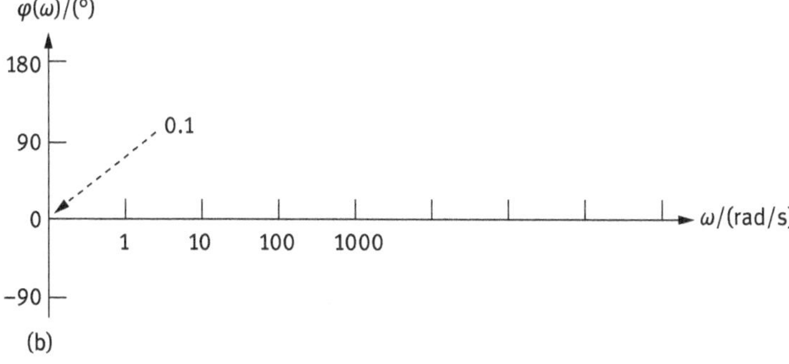

Fig. 7.12: The Bode diagram: (a) logarithm amplitude characteristic diagram and (b) logarithm phase characteristic diagram.

In Fig. 7.12(a), the Y-coordinate represents the decibel of the amplitude, that is, $L(\omega) = 20\lg |G(j\omega)|$, with the linear unit of dB. The X-coordinate represents the value of ω with the unit of rad/s. In Fig.7.12(b), the Y-coordinate represents the phase of $G(j\omega)$ with the linear unit of degree. (The phase of two vectors should be added when those two vectors multiply.) The X-coordinate represents the value of ω with the unit of rad/s.

7.2.3.1 Coordinate Dividing

To further understand the logarithm amplitude characteristic diagram, we introduce Fig. 7.13, in which linear dividing with the decibel unit of $L(\omega)$ is used in the Y-coordinate, and logarithm dividing ($\lg\omega$) but labeled with ω is used in the X-coordinate. The purpose of the coordinate dividing is for simply plotting and compacting the whole logarithm amplitude characteristic diagram.

7.2.3.2 Slope of the Asymptote

In fact, here we plot an asymptote of the logarithm amplitude characteristic diagram, which consists of some line segments. An asymptote is enough for plotting logarithm

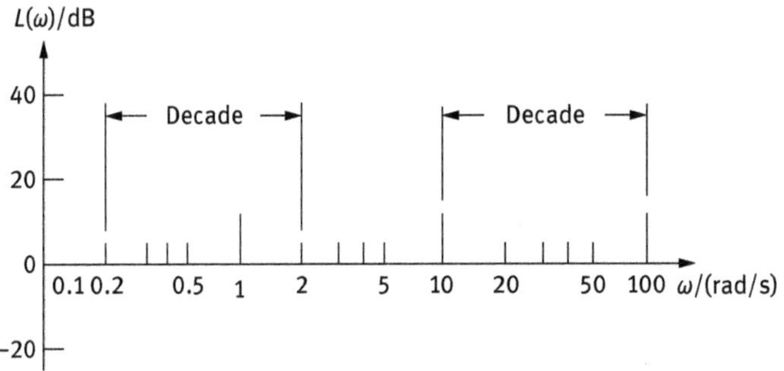

Fig. 7.13: The coordinate division for logarithm amplitude characteristic diagram.

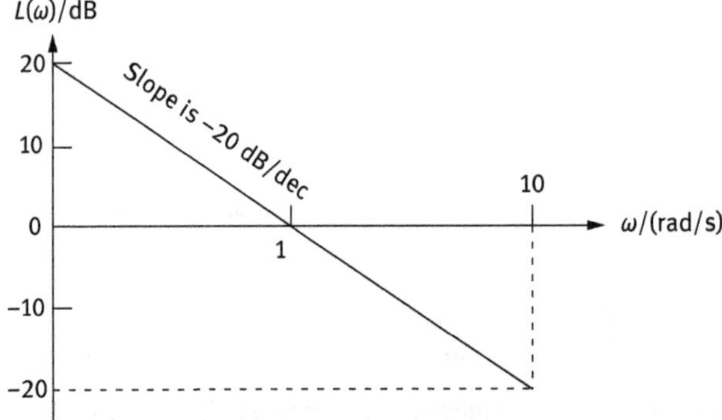

Fig. 7.14: The slope for logarithm amplitude asymptote.

amplitude diagram $L(\omega)$, and so first the slope of the asymptote is required. The changed decibel of $L(\omega)$ when frequency ω increases 10 times is called the slope of the asymptote. For example, in Fig. 7.13, if $\omega_1 = 10$ and $\omega_2 = 100$, the distance between ω_1 and ω_2 is decade (dec) for abbreviation. If frequency increases 10 times (decade) and $L(\omega)$ decays 20 decibel, the slope of the asymptote is given as –20 dB/dec, as shown in Fig. 7.14.

In Fig. 7.14, the logarithm amplitude characteristic of the link is $L(\omega) = -20\lg\omega$. When ω changes from ω_1 to $10\,\omega_1$, the amplitude values for the two frequency, ω_1 and ω_2, are as follows:

$$L(\omega_1) = -20\lg\omega_1$$

$$L(10\omega_1) = -20\lg 10\omega_1 = -20\lg\omega_1 - 20$$

Thus, when frequency ω increases 10 times, the amplitude values decay,

$$L(10\omega_1) - L(\omega_1) = -20$$

From the above equation, one can conclude that when frequency changes from ω_1 to 10 ω_1, the frequency increases 10 times (dec). In another words, when $L(\omega)$ changes from ($-20\lg\omega_1$) to ($-20\lg\omega_1 -20$), the amplitude values decay 20 decibels (dB). So, the slope of the asymptote is −20 dB/dec.

Advantages of designing logarithm coordinates are as follows:

(1) The logarithm amplitude characteristic diagram expressed as a product of more than one term can be obtained by adding the individual dB amplitude diagrams for each product term. Owing to the logarithm coordinate, a multiplying – dividing operation is turned into an adding –subtracting operation.

(2) We consider the Bode diagrams as semi-logarithmic coordinate, because its *X*-coordinate is divided by logarithm while its *Y*-coordinate is divided linearly.

① When compared with linear division, the logarithm division in *X*-coordinate is so convenient that we can find out about the frequency characteristics in a wide frequency range of the *X*-coordinate.

② The unit of *Y*-coordinate is decibel, which can reduce the value of slope of the logarithm amplitude characteristic. For example, if $|G(\mathrm{j}\omega)|$ changes from 1 to 1,000, 20lg $|G(\mathrm{j}\omega)|$ changes from 20lg1000 to 20lg1, that is, 20lg1000 − 20lg1 = 60 dB. In other words, the range of the *Y*-coordinate becomes narrow.

7.2.3.3 The Asymptote of the Logarithm Amplitude Characteristic Curve

Let amplitude characteristic of the link be

$$|G(\mathrm{j}\omega)| = \frac{5}{\sqrt{1+T^2\omega^2}}$$

The logarithm amplitude characteristic is

$$L(\omega) = 20\lg|G(\mathrm{j}\omega)| = 20\left(\lg 5 - \lg\sqrt{1+T^2\omega^2}\right)$$

There are two steps for plotting the asymptote of $L(\omega)$:

(1) $T\omega \ll 1$.

$$L(\omega) = 20\left(\lg 5 - \lg\sqrt{1+T^2\omega^2}\right) = 20\left(\lg 5 - \lg\sqrt{1}\right) = 20\lg 5 = 14 \text{ dB}$$

Hence, this is a straight line, with a 14-decibel distance to *X*-coordinate and a slope of 0 dB/dec.

(2) $T\omega \gg 1$.

$$L(\omega) = 20\left(\lg 5 - \lg\sqrt{1+T^2\omega^2}\right) = 20\lg 5 - 20\lg T\omega = 20\lg\frac{5}{T} - 20\lg\omega$$

This is an oblique line, with a slope of −20 dB/dec.

Thus, the logarithm amplitude characteristic curve of this link can be represented by the above two lines: $L(\omega) = 20\lg 5$ and $L(\omega) = 20\lg\frac{5}{T} - 20\lg\omega$.

These two lines are asymptotes for the logarithm amplitude characteristic curve of this link.

7.2.3.4 Break Frequency

Using the same example as in Section 7.2.3.3, the frequency at which two asymptotes intersect is called the break frequency (ω_T). We can solve two asymptotes to obtain the break frequency:

$$\begin{cases} \text{Low frequency}: & L(\omega) = 20\lg 5 \\ \text{High frequency}: & L(\omega) = 20\lg 5 - 20\lg T\omega \end{cases}$$

Hence, the break frequency is

$$\omega_T = \frac{1}{T}$$

The break frequency is the reciprocal of the time constant or undamped natural frequency. The slope of the asymptote will occur break or abrupt change at break frequency. So, it is very important to first ensure every break frequency before we plot the Bode diagram for the whole system.

7.2.3.5 Amplitude Crossing Frequency

Using the same example as in Section 7.2.3.3, the frequency at which the logarithm amplitude characteristic curve and X-coordinate intersect with each other is called the amplitude crossing frequency (ω_c). We can solve the high-frequency asymptote equation and $L(\omega) = 0$ to get ω_c,

$$\begin{cases} L(\omega) = 0 \\ L(\omega) = 20\lg 5 - 20\lg T\omega \end{cases}$$

Thus, amplitude crossing frequency is

$$\omega_c = \frac{5}{T}$$

7.2.3.6 Phase Crossing Frequency

The frequency where logarithm phase characteristic curve and the line of −180° intersect is called the phase crossing frequency (ω_g),

$$\angle G(j\omega_g) = -180^\circ$$

7.2.4 The Bode Diagrams for Typical Links

7.2.4.1 Proportion Link

The TF of the proportion link is

$$G(s) = K$$

Therefore, the frequency characteristic function is

$$G(j\omega) = K \tag{7.23}$$

Thus, the logarithm amplitude characteristic of the proportion link is

$$20\lg|G(j\omega)| = 20\lg K \tag{7.24}$$

The logarithm phase characteristic of the proportion link is

$$\varphi(\omega) = \angle G(j\omega) = 0^\circ \tag{7.25}$$

As shown in Fig. 7.15(a), the logarithm amplitude characteristic curve is a horizontal line, with the distance $20\lg K$ (dB) to the X-axis. The logarithm phase characteristic curve is a horizontal line, which actually is the horizontal axis, as shown in Fig. 7.15(b).

As shown in Fig. 7.15, when K changes, the logarithm amplitude characteristic curve will move up or down in a vertical direction but the logarithm phase characteristic curve will not. The system gain of several links in series is equal to the product of gains of those links, that is, $K = K_1 \cdot K_2 \ldots K_n$. The logarithm amplitude characteristic of the system is equal to the sum of logarithm gain of each link:

$$\begin{aligned} L(\omega) &= 20\lg K = 20 lg K_1 K_2 \cdots K_n = 20(\lg K_1 + \lg K_2 + \cdots + \lg K_n) \\ &= L(\omega_1) + L(\omega_2) + \cdots + L(\omega_n) \end{aligned}$$

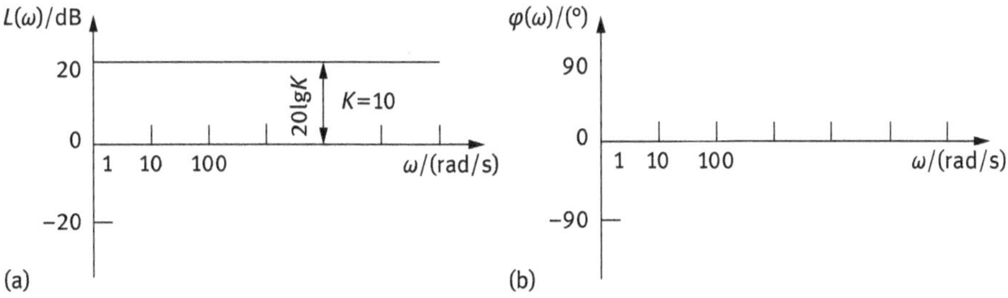

Fig. 7.15: The Bode diagrams for the proportion link: (a) the logarithm amplitude characteristic diagram and (b) the logarithm phase characteristic diagram.

7.2.4.2 Integral Link

The TF of the integral link is

$$G(s)=\frac{1}{s}$$

So, the frequency characteristic function is

$$G(\mathrm{j}\omega)=\frac{1}{\mathrm{j}\omega}=-\mathrm{j}\frac{1}{\omega} \tag{7.26}$$

The logarithm amplitude characteristic of the integral link is

$$20\lg|G(\mathrm{j}\omega)|=20\lg\frac{1}{\omega}=-20\lg\omega \tag{7.27}$$

The logarithm phase characteristic of the integral link is

$$\varphi(\omega)=\angle G(\mathrm{j}\omega)=-90^{\circ} \tag{7.28}$$

According to eq. (7.27), when $\omega = 0.1$ rad/s, $20\lg|G(\mathrm{j}\omega)| = -20\lg 0.1 = 20$ dB. It is obvious that point A(0.1, 20) is on the logarithm amplitude characteristic curve. When $\omega = 1$ rad/s, $20\lg|G(\mathrm{j}\omega)| = -20\lg 1 = 0$ dB. This is the point B(1, 0) on the logarithm amplitude characteristic curve. (In fact, $\omega = 1$ rad/s is the amplitude crossing frequency ω_{c}.) When $\omega = 10$ rad/s, $20\lg|G(\mathrm{j}\omega)| = -20\lg 10 = -20$ dB. This is the point C(10, −20) on the logarithm amplitude characteristic curve, as shown in Fig. 7.16(a).

In terms of these three special points, the logarithm amplitude characteristic curve will decay at 20 dB when the frequency increases 10 times. The logarithm amplitude characteristic curve is an oblique line, which passes through the point (1, 0), with the slope of −20 dB/dec. According to eq. (7.28), the logarithm phase characteristic curve of the integral link is a straight line parallel to the horizontal axis, which passes the point (0, −90°) and has no relationship with the frequency ω, as shown in Fig. 7.16 (b).

Here is a useful further extension about plotting the Bode diagram for several integral links. The TF of several integral links is

$$G(s)=\frac{1}{s^{N}}$$

The frequency characteristic function is

$$G(\mathrm{j}\omega)=\frac{1}{(\mathrm{j}\omega)^{N}}$$

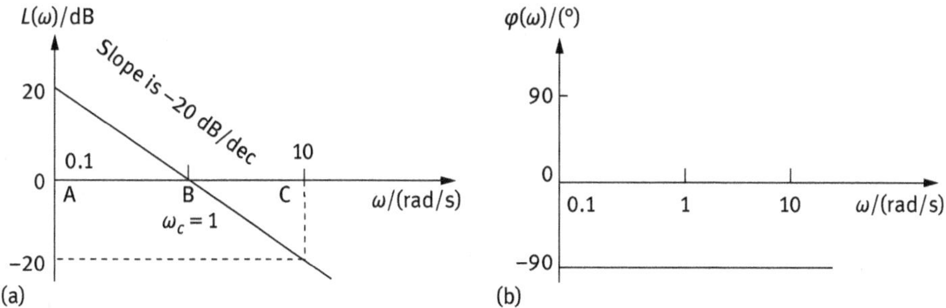

Fig. 7.16: The Bode diagrams for the integral link: (a) the logarithm amplitude characteristic diagram and (b) the logarithm phase characteristic diagram.

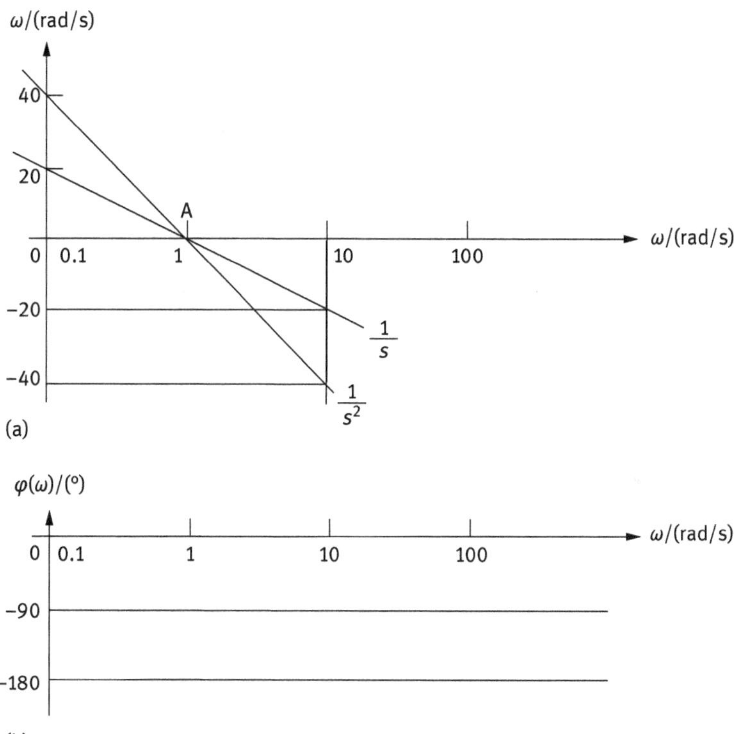

Fig. 7.17: The Bode diagrams for the integral links in series: (a) the logarithm amplitude characteristic diagram and (b) the logarithm phase characteristic diagram.

The logarithm amplitude characteristic of several integral links is

$$20\lg|G(\mathrm{j}\omega)| = 20\lg\frac{1}{\omega^N} = -20N\lg\omega$$

The logarithm phase characteristic of several integral links is

$$\varphi(\omega) = \angle G(j\omega) = -90°N$$

When $\omega = 0.1$ rad/s, we find that $20\lg|G(j\omega)| = -20N\lg 0.1 = 20N$ dB. Thus, the point (0.1, 20N) is on the logarithm amplitude characteristic curve. When $\omega = 1$ rad/s, we find that $20\lg|G(j\omega)| = -20N\lg 1 = 0$ dB. This is the point A(1, 0) on the logarithm amplitude characteristic curve, shown in Fig. 7.16 (a). Here, $\omega = 1$ rad/s is the amplitude of crossing frequency ω_c. When $\omega = 10$ rad/s, we find that $20\lg|G(j\omega)| = -20N\lg 10 = -20N$ dB. That is the point (10, $-20N$) on the logarithm amplitude characteristic curve.

On the basis of the aforementioned analysis, we can conclude that the logarithm amplitude characteristic curve of several integral links is an oblique line, which passes the point A(1, 0) with the slope of $-20N$ dB/dec. Its logarithm phase characteristic curve is a straight line parallel to the horizontal axis, which passes the point (0, $-90°N$). From Fig. 7.17(a), for example, if $N = 2$, the logarithm amplitude characteristic curve of two integral links is an oblique line, which passes the point A(1, 0) with the slope of −40 dB/dec. Its logarithm phase characteristic curve is a straight line parallel to the horizontal axis, which passes the point (0, −180°), as shown in Fig. 7.17(b).

Example 7.5
The TF of a control system is given by

$$G(s) = \frac{10}{s^2}$$

Plot the Bode diagram.

Solution
The frequency characteristic function is

$$G(j\omega) = \frac{10}{-\omega^2} \tag{7.29}$$

Thus, the logarithm amplitude characteristic of this system is

$$L(\omega) = 20\lg|G(j\omega)| = 20\lg\frac{10}{\omega^2} = 20 - 40\lg\omega \tag{7.30}$$

The logarithm phase characteristic of this system is

$$\varphi(\omega) = \angle G(j\omega) = -180° \tag{7.31}$$

Using the same method of analysis, we can use three special points, A(0.1, 60), B(1, 20), and C(10, −20), to find that the slope of the logarithm amplitude characteristic curve is −40 dB/dec, as shown in Fig. 7.18(a). According to eq. (7.31), the logarithm phase characteristic curve of the integral link is a straight line parallel to the horizontal axis, which passes the point (0, −180°) and has no relationship with the frequency ω, as shown in Fig. 7.18 (b).

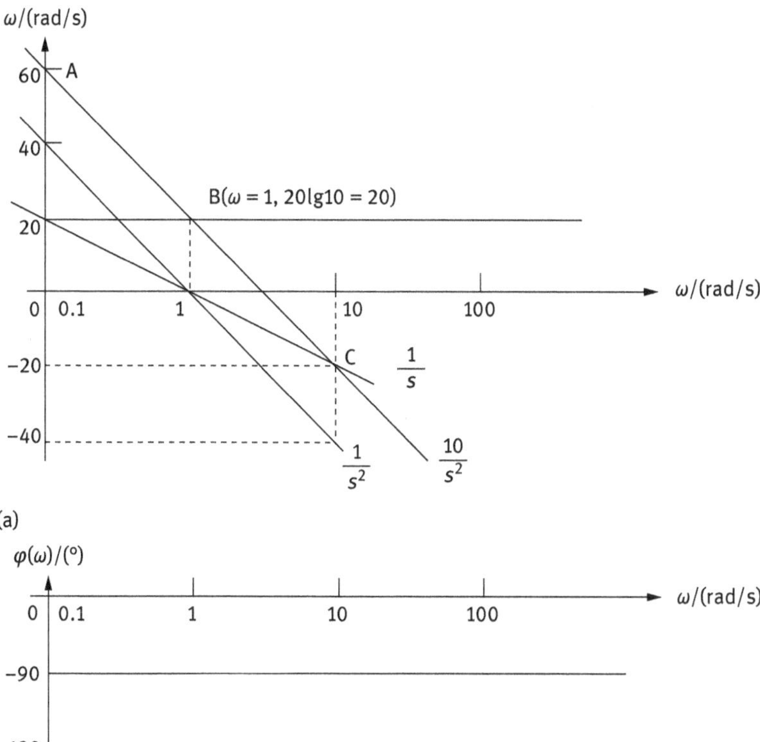

Fig. 7.18: The Bode diagrams for two integral links and proportion link: (a) the logarithm amplitude characteristic diagram and (b) the logarithm phase characteristic diagram.

Now, we will discuss some useful tips for plotting integral links and proportional link,

(1) Point (1, 0) must be on the logarithm amplitude characteristic curve of the integral link. That is, when $\omega = 1$, $L(\omega) = -20\lg 1 = 0$. Thus, the logarithm amplitude characteristic curve of the single integral link should pass the point (1, 0) with the slope of −20 dB/dec.

(2) The slope of the asymptote of N integral links $1/s^N$ is $-20N$ dB/dec. Hence, the logarithm amplitude characteristic curve of $1/s^N$ should pass the point (1, 0) with the slope of $-20N$ dB/dec.

(3) For the gain in the numerator, the logarithm amplitude characteristic curve of K/s^N should pass the point $(1, 20\lg K)$ with the slope of $-20N$ dB/dec. In other words, the logarithm amplitude characteristic curve of K/s^N is the one that $1/s^N$ is translated $20\lg K$(dB) along the vertical axis.

(4) The logarithm phase characteristic curve of K/s^N is a horizontal line with the vertical distance of $-90°N$ to the X-coordinate.

So, here is an important conclusion. In Fig. 7.19, according to the two-point formula of the line using the points A and B, the line slope is

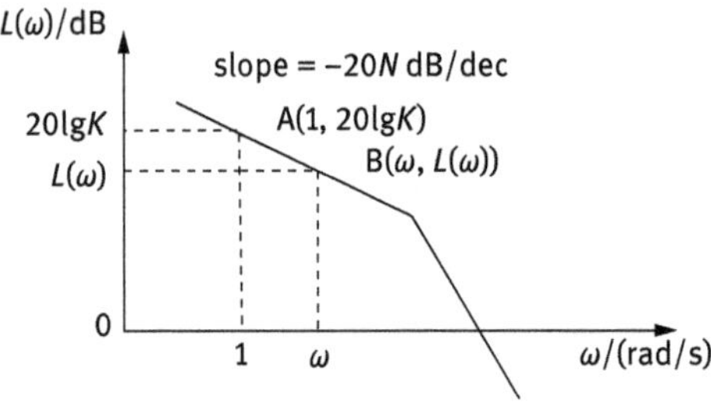

Fig. 7.19: The logarithm amplitude characteristic curve for the low frequency.

$$k = \frac{y_1 - y_2}{x_1 - x_2}$$

Thus, for the points A and B in Fig. 7.19, we have

$$-20N = \frac{20\lg K - L(\omega)}{\lg 1 - \lg \omega}$$

That is,

$$20\lg \frac{K}{\omega^N} = L(\omega) \tag{7.32}$$

If the low-frequency TF of the system is composed of the integral and proportional links, eq. (7.32) could be used as an equation. In terms of this equation, we could get the amplitude value for any point existing in the low-frequency curve.

7.2.4.3 Ideal Differential Link

The TF of the ideal differential link is

$$G(s) = s$$

Thus, the frequency characteristic function is

$$G(\mathrm{j}\omega) = \mathrm{j}\omega \tag{7.33}$$

The logarithm amplitude characteristic of the ideal differential link is

$$20\lg|G(\mathrm{j}\omega)| = 20\lg\omega \tag{7.34}$$

The logarithm phase characteristic of the ideal differential link is

$$\varphi(\omega) = \angle G(\mathrm{j}\omega) = 90^\circ \tag{7.35}$$

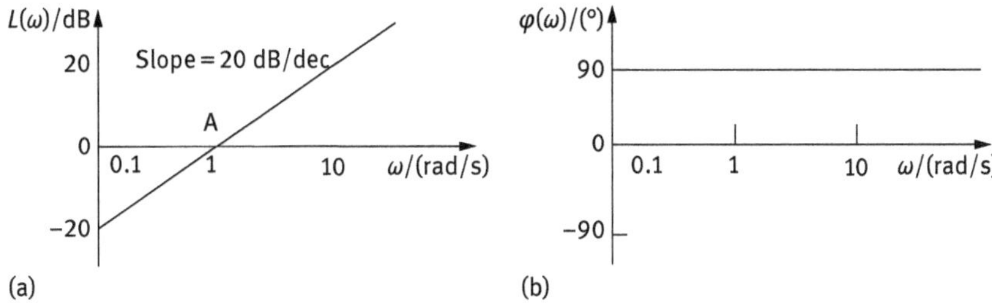

Fig. 7.20: The Bode diagrams for the ideal differential link: (a) the logarithm amplitude characteristic diagram and (b) the logarithm phase characteristic diagram.

By means of eq. (7.34), when $\omega = 0.1$, the logarithm amplitude characteristic of the ideal differential link is $20\lg|G(\mathrm{j}\omega)| = -20$ dB, and when $\omega = 1$, $20\lg|G(\mathrm{j}\omega)| = 0$ dB. In Fig. 7.20, the logarithm amplitude characteristic curve of the ideal differential link is an oblique line, which passes the point A(1, 0) with the slope of 20 dB/dec. The logarithm phase characteristic curve of the ideal differential link is a straight line parallel to the horizontal axis, which passes the point (0, 90°) and has no relationship with the frequency ω.

7.2.4.4 Inertial Link

The TF of the inertial link is

$$G(s) = \frac{1}{Ts+1}$$

Thus, the frequency characteristic function is

$$G(\mathrm{j}\omega) = \frac{1}{1+\omega^2T^2} - \frac{T\omega}{1+\omega^2T^2}\cdot \mathrm{j} \tag{7.36}$$

The amplitude characteristic of the inertial link is

$$|G(\mathrm{j}\omega)| = \sqrt{\left(\frac{1}{1+T^2\omega^2}\right)^2 + \left(\frac{T\omega}{1+T^2\omega^2}\right)^2} = \frac{1}{\sqrt{1+T^2\omega^2}} \tag{7.37a}$$

The logarithm amplitude characteristic of the inertial link is

$$L(\omega) = 20\lg|G(\mathrm{j}\omega)| = -20\lg\sqrt{1+T^2\omega^2} \tag{7.37b}$$

And the logarithm phase characteristic of the integral link is

$$\varphi(\omega) = \angle G(\mathrm{j}\omega) = -\arctan T\omega \tag{7.38}$$

When $\omega \ll 1/T$, it is a low-frequency curve; eq. (7.37b) could be changed to

$$L(\omega) = -20\lg\sqrt{1+T^2\omega^2} \approx -20\lg 1 \approx 0\ \text{dB} \tag{7.39}$$

This is a straight line, which coincides with the X-axis. In other words, the asymptote is a horizontal line coinciding with the X-axis.

When $\omega \gg 1/T$, it is a high-frequency curve; eq. (7.37b) could be changed to

$$L(\omega) = -20\lg\sqrt{1+T^2\omega^2} \approx -20\lg\omega T \tag{7.40}$$

This is an oblique line with the slope of −20 dB/dec. In other words, the asymptote is an oblique line with the slope of −20 dB/dec.

On the basis of the aforementioned analysis, the intersection frequency of two asymptotes is the break frequency ω_T

$$\begin{cases} L(\omega) = 20\lg|G(\mathrm{j}\omega)| \approx 0\ \text{dB} \\ L(\omega) = 20\lg|G(\mathrm{j}\omega)| \approx -20\lg\omega T \end{cases}$$

Solving the above set of equations to get the value for ω_T,

$$\omega_T = \frac{1}{T}$$

Here, we could get the amplitude crossing frequency ω_c based on the following set of equations:

$$\begin{cases} L(\omega) = 0\ \text{dB} \\ L(\omega) = 20\lg|G(\mathrm{j}\omega)| \approx -20\lg\omega T \end{cases}$$

Hence, the amplitude crossing frequency ω_c is

$$\omega_C = \frac{1}{T}$$

Hence, for the inertial link,

$$\omega_T = \omega_C = \frac{1}{T}$$

Now, here are a few steps about how to plot Bode diagrams for the inertial link.

Step 1. Plotting the asymptote of $L(\omega)$

From eqs. (7.39) and (7.40), we can get the asymptote of $L(\omega)$ as shown in Fig. 7.21 (a). When $\omega = \omega_T$, the low-frequency asymptote and the high-frequency asymptote intersect at a point $(\omega_T, 0)$.

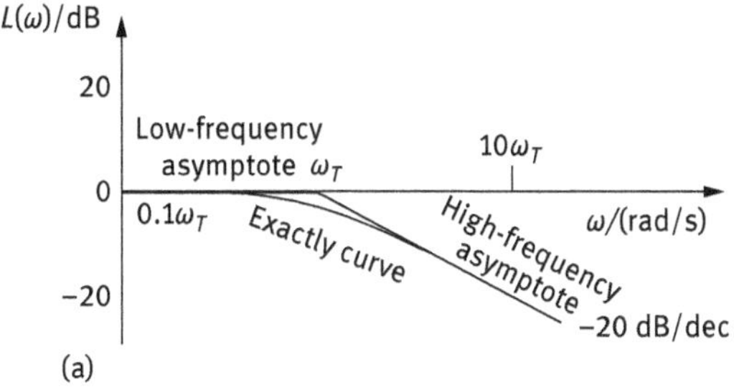

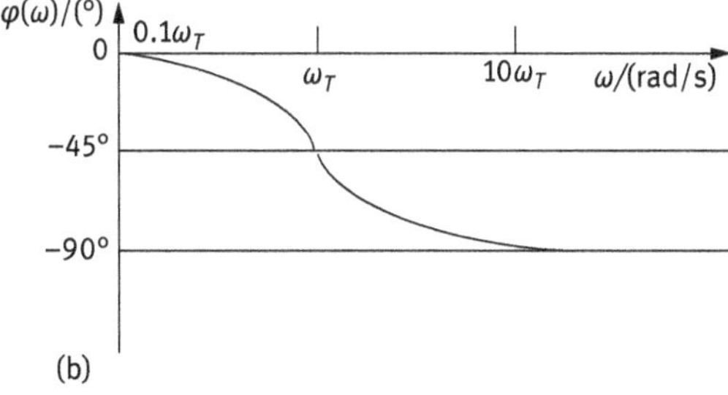

Fig. 7.21: The Bode diagrams for the inertial link: (a) the logarithm amplitude characteristic diagram and (b) the logarithm phase characteristic diagram.

Step 2. Plotting the logarithm phase characteristic curve

To obtain the exact logarithm phase characteristic curve, we should calculate every $\varphi(\omega)$ in every ω using eq. (7.38), as listed in Table 7.1. Then, based on those points, we can plot the logarithm phase characteristic curve, as shown in Fig. 7.21(b). When $\omega = \omega_T$, the error between the exact curve and the asymptote would be

$$e = 20\lg|G(\mathrm{j}\omega)||_{\omega=\omega_T} - 0 = 20\lg|G(\mathrm{j}\omega_T)| - 0 = -20\lg\sqrt{1+T^2\frac{1}{T^2}} \approx -3$$

The error value is shown in Fig. 7.22.

7.2.4.5 First-Order Differential Link

The TF of the first-order differential link is

$$G(s) = Ts + 1$$

Hence, the frequency characteristic function is

$$G(\mathrm{j}\omega) = 1 + T\omega\mathrm{j} \tag{7.41}$$

The amplitude characteristic of the first-order differential link is

$$|G(\mathrm{j}\omega)| = \sqrt{1+T^2\omega^2} \tag{7.42a}$$

Table 7.1: The values of $\varphi(\omega)$ in every ω.

Ω	$\frac{1}{10T}$	$\frac{1}{5T}$	$\frac{1}{2T}$	$\frac{1}{T}$	$\frac{2}{T}$	$\frac{5}{T}$	$\frac{10}{T}$
$\varphi(\omega)/(°)$	−5.7	−11.3	−26.2	−45	−63.4	−78.7	−84.3

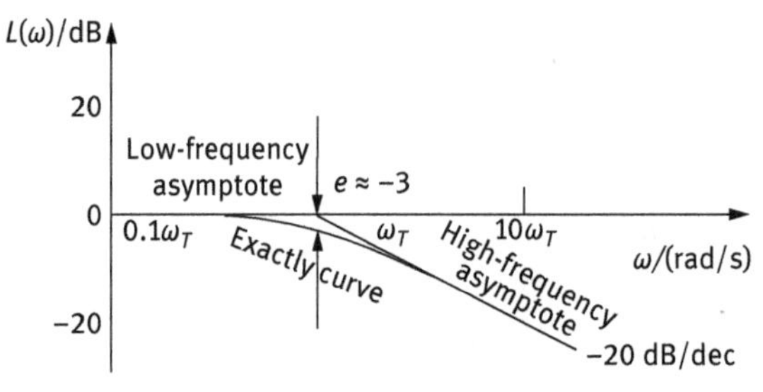

Fig. 7.22: The error between the exact curve and the asymptote for the inertial link.

The logarithm amplitude characteristic of the first-order differential link is

$$L(\omega) = 20\lg|G(j\omega)| = 20\lg\sqrt{1+T^2\omega^2} \tag{7.42b}$$

And the logarithm phase characteristic of the first-order differential link is

$$\varphi(\omega) = \angle G(j\omega) = \arctan T\omega \tag{7.43}$$

When $\omega \ll 1/T$, it is a low-frequency curve; eq. (7.42b) could be similar to the following formula:

$$L(\omega) = 20\lg\sqrt{1+T^2\omega^2} \approx 20\lg 1 \approx 0\text{dB} \tag{7.44}$$

When $\omega \gg 1/T$, it is a high-frequency curve; eq. (7.42b) changes to

$$L(\omega) = 20\lg\sqrt{1+T^2\omega^2} \approx 20\lg\omega T \tag{7.45}$$

With the same method of analysis, we could calculate every $\varphi(\omega)$ in every ω using eq. (7.43). On the basis of those points $(\omega, \varphi(\omega))$, we can draw the logarithm phase characteristic diagram for the first-order differential link, as shown in Fig. 7.23.

7.2.4.6 Second-Order Oscillation Link

The TF of the second-order oscillation link is

$$G(s) = \frac{1}{T^2s^2 + 2\zeta Ts + 1}$$

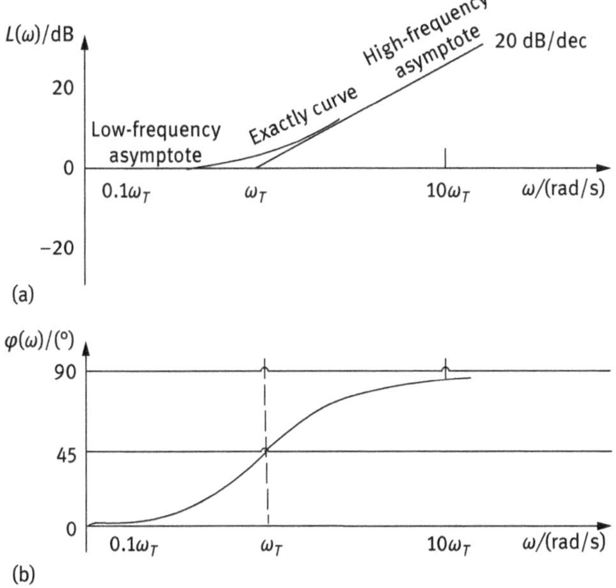

Fig. 7.23: The Bode diagrams for the first-order differential link: (a) the logarithm amplitude characteristic diagram and (b) the logarithm phase characteristic diagram.

The frequency characteristic function is

$$G(\mathrm{j}\omega) = \frac{1 - T^2\omega^2}{(1 - T^2\omega^2)^2 + (2\zeta T\omega)^2} - \mathrm{j}\frac{2\zeta T\omega}{(1 - T^2\omega^2)^2 + (2\zeta T\omega)^2} \tag{7.46}$$

The amplitude characteristic of the second-order oscillation link is

$$|G(\mathrm{j}\omega)| = \frac{1}{\sqrt{(1 - T^2\omega^2)^2 + (2\zeta T\omega)^2}} \tag{7.47a}$$

The logarithm amplitude characteristic of the second-order oscillation link is

$$L(\omega) = 20\lg|G(\mathrm{j}\omega)| = -20\lg\sqrt{(1 - T^2\omega^2)^2 + (2\zeta T\omega)^2} \tag{7.47b}$$

And the logarithm phase characteristic of the integral link is

$$\varphi(\omega) = \angle G(\mathrm{j}\omega) = -\arctan\frac{2\zeta T\omega}{1 - T^2\omega^2} \tag{7.48}$$

The logarithm amplitude and phase characteristics of the second-order oscillation link are related not only with ω but also with the damping ratio ζ. We should omit damping ratio ζ to get the asymptote, and then consider ζ to mend the asymptote.

According to eq. (7.47b), when $\omega \ll 1/T$, it is a low-frequency curve; eq. (7.47b) is changed to

$$L(\omega) = -20\lg\sqrt{(1-T^2\omega^2)^2} \approx 0 \text{ dB} \tag{7.49}$$

Hence, it is a straight line, which coincides with the X-axis.

When $\omega \gg 1/T$, it is a high-frequency curve; eq. (7.47b) is changed to

$$L(\omega) = -20\lg\sqrt{(1-T^2\omega^2)^2} \approx -40\lg\omega T \tag{7.50}$$

Therefore, it is an oblique line with the slope of –40 dB/dec.

Because the low-frequency curve is just the X-axis, the break frequency ω_c and the amplitude crossing frequency ω_T are the same, and the logarithm amplitude frequency characteristic diagram is shown in Fig. 7.24 (a).

From eq. (7.48), we can find that the variation range of $\varphi(\omega)$ is (0°, 180°). To plot the $\varphi(\omega)$ diagram, we must adopt the point depiction method. In other words, we must plot $\varphi(\omega)$ diagram according to the every point value for $(\omega, \varphi(\omega))$. The logarithm phase frequency characteristic diagram is shown in Fig. 7.24(b).

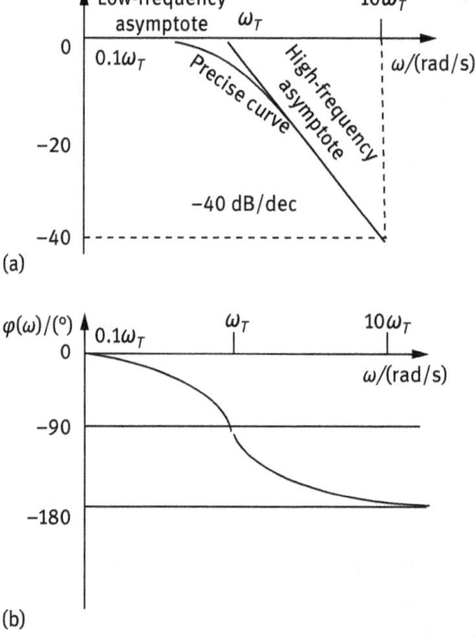

Fig. 7.24: The Bode diagrams for the second-order oscillation link: (a) the logarithm amplitude characteristic diagram and (b) the logarithm phase characteristic diagram.

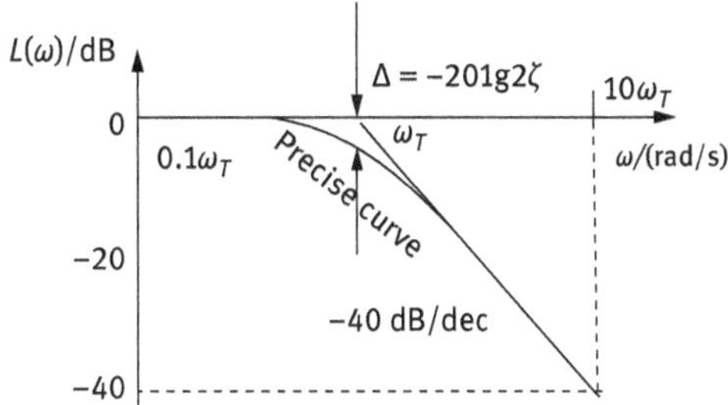

Fig. 7.25: The error between the exact curve and the asymptote for the second-order oscillation link.

As mentioned earlier, to obtain the asymptote, we omit damping ratio ζ at the beginning of evaluation. As shown in Fig. 7.25, when $\omega = 1/T$, the error of the precise logarithm amplitude characteristic and the asymptote is

$$\begin{aligned}\Delta &= -20\lg\sqrt{(1-T^2\omega^2)^2+(2\zeta T\omega)^2}-0\\ &= -20\lg\sqrt{\left(1-T^2\frac{1}{T^2}\right)^2+\left(2\zeta T\frac{1}{T}\right)^2} = -20\lg 2\zeta\end{aligned}$$

7.2.4.7 Second-Order Differential Link

The TF of the second-order differential link is

$$G(s) = T^2s^2 + 2\zeta Ts + 1$$

The frequency characteristic function is

$$G(\mathrm{j}\omega) = T^2(\mathrm{j}\omega)^2 + 2\zeta T(\mathrm{j}\omega) + 1 \tag{7.51}$$

The logarithm amplitude characteristic of the second-order differential link is

$$L(\omega) = 20\lg\sqrt{(1-T^2\omega^2)^2+(2\zeta T\omega)^2} \tag{7.52}$$

And the logarithm phase characteristic of the second-order differential link is

$$\varphi(\omega) = \arctan\frac{2\zeta T\omega}{1-T^2\omega^2} \tag{7.53}$$

The Bode diagram of the second-order differential link is shown in Fig. 7.26. From eqs. (7.47b) and (7.52), we can find that the logarithm amplitude characteristic of the second-order oscillation link and the second-order differential link has a map relationship. From eqs. (7.48) and (7.53), we can find the same map relationship between their logarithm phase frequency characteristic diagrams. The difference between second-order differential link and second-order oscillation link about their logarithm characteristics is only the signal, minus.

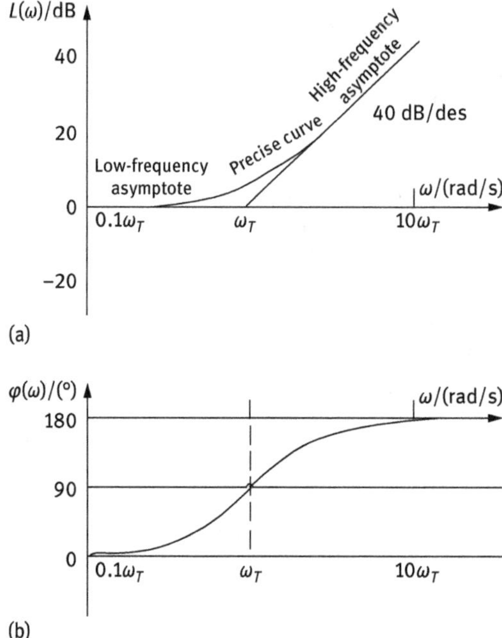

Fig. 7.26: The Bode diagrams for the second-order differential link: (a) the logarithm amplitude characteristic diagram and (b) the logarithm phase characteristic diagram.

7.2.4.8 Summary

Figure 7.27 gives Bode diagrams for the seven kinds of typical links. From Fig. 7.27, we can find some map relationships among those links. From Fig. 7.27, asymptote nos ① and ② are the ideal differential link and the integral link. Asymptote nos ③ and ⑥ are the second-order differential link and the second-order oscillation link. Asymptote Nos ④ and ⑤ are the first-order differential link and the inertial link. Asymptote no. ⑦ is the proportion link with $K = 10$. The map relationships of the logarithmic characteristic asymptotes would help us to correctly remember the Bode diagrams for different links.

7.3 The Open-Loop Bode Diagram of Control System

Why did we plot so many Bode diagrams for so many typical links? What are Bode diagrams useful for? The answers are plotting the open-loop Bode diagrams of control system and getting the closed-loop characteristics by means of the open-loop Bode diagrams. A significant amount of work is required for directly plotting a open-loop Bode diagram. Fortunately, the open-loop diagram is composed of several typical links. Thus, with the help of the frequency characteristic of the typical links, we can easily plot the open-loop Bode diagram. The general form of the open-loop TF for control system is

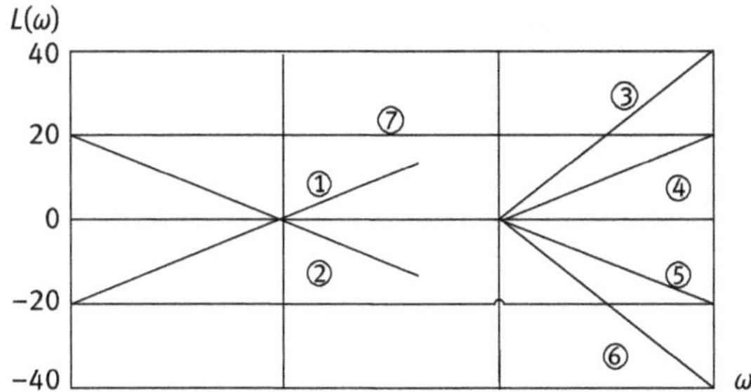

(a)

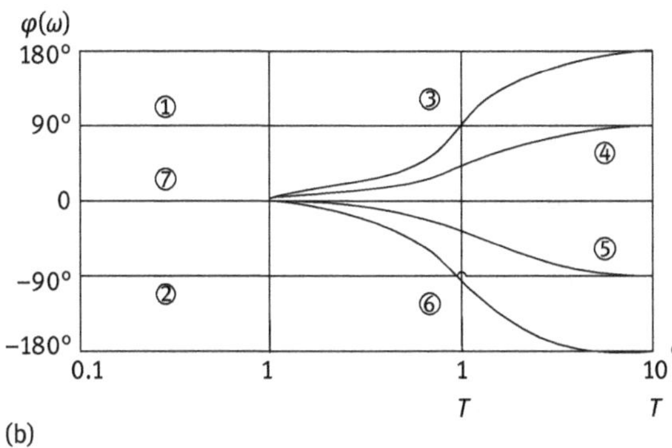

(b)

Fig. 7.27: The map relationship of the logarithm characteristic asymptotes: (a) the logarithm amplitude characteristic diagram for every link and (b) the logarithm phase characteristic diagram for every link.

$$G(s)H(s)=\frac{K\prod_{i=1}^{m}(\tau_i s+1)\prod_{l=1}^{n}(\tau_l^2 s^2+2\zeta_l\tau_l s+1)}{s^N\prod_{j=1}^{p}(T_j s+1)\prod_{k=1}^{q}(T_k^2 s^2+2\zeta_k T_k s+1)} \tag{7.54}$$

where N is the number of the integral link; p is the number of the inertial link; q is the number of the second-order oscillation link; m is the number of the first-order differential link; and n is the number of the second-order differential link.

Thus, from eq. (7.54), the open-loop TF is composed of every link in series. And the logarithm amplitude frequency characteristic is

$$\begin{aligned}L(\omega)=&20\lg K-N20\lg\omega-\sum_{j=1}^{p}20\lg\sqrt{1+T_j^2\omega^2}\\&-\sum_{k=1}^{q}20\lg\sqrt{(1-T_k^2\omega^2)^2+(2\zeta_k T_k\omega)^2}\\&+\sum_{i=1}^{m}20\lg\sqrt{1+\tau_i^2\omega^2}+\sum_{l=1}^{n}20\lg\sqrt{(1-\tau_l^2\omega^2)^2+(2\zeta_l\tau_l\omega)^2}\end{aligned} \tag{7.55}$$

The logarithm phase frequency characteristic is

$$\begin{aligned}\varphi(\omega) = -N\tfrac{\pi}{2} - \sum_{j=1}^{p} \arctan T_j\omega - \sum_{k=1}^{q} \arctan \frac{2\zeta_k T_k \omega}{1 - T_k^2\omega^2} \\ + \sum_{i=1}^{m} \arctan \tau_i\omega + \sum_{l=1}^{n} \arctan \frac{2\zeta_l \tau_l \omega}{1 - \tau_l^2\omega^2}\end{aligned} \tag{7.56}$$

Here are some steps for plotting an open-loop Bode diagram:

(1) Transfer an open-loop TF into a time constant form, as shown in eq. (7.54).

(2) Confirm the proper scale for coordinates, especially for *X* coordinate.

(3) Find the break frequency ω_T of the inertial link, first-order differential link, second-order oscillation link, and second-order differential link and plot them on the frequency axis from smaller to bigger number.

(4) Evaluate 20lg*K* (dB) according to the open-loop gain *K*. In fact, one can look upon the open-loop gain as the proportion link.

(5) For the low-frequency asymptote,

① If there is only the proportion link, one should plot a horizontal line with the amplitude value 20lg*K*.

② If there is only the integral link, one can draw an oblique line passing point (1, 0) with the slope of $-20N$ dB/dec.

③ If proportion link and integral link both exist in the low-frequency asymptote, one should plot an oblique line passing point (ω =1 rad/s, 20lg*K*) with the slope of $-20N$ dB/dec.

(6) For the high-frequency asymptote, one should change the slope of the asymptote in every break frequency ω_T for every link.

① For the inertial link, change the slope of the asymptote in the break frequency ω_T to $-20p$ dB/dec.

② For the second-order oscillation link, change the slope of the asymptote in the break frequency ω_T to $-40q$ dB/dec.

③ For the first-order differential link, change the slope of the asymptote in the break frequency ω_T to +20 *m*dB/dec.

④ For the second-differential link, change the slope of the asymptote in the break frequency ω_T to $+40n$ dB/dec.

(7) Compensation the asymptote to get the precise logarithm amplitude characteristic diagram if necessary.

(8) Plot the logarithm phase characteristic diagrams for every link and superpose them in the same frequency to get the final open-loop logarithm phase characteristic diagram.

Example 7.6

Consider the open-loop TF of the control system,

$$G(s) = \frac{10(s+3)}{s(s+2)(s^2+s+2)}$$

Plot the open-loop Bode diagrams for this control system.

Solution

Changing the general form into the time constant form, also called as "+1" form, we get

$$G(s) = \frac{7.5\left(\frac{s}{3}+1\right)}{s\left(\frac{1}{2}s+1\right)\left[\left(\frac{s}{\sqrt{2}}\right)^2 + \frac{s}{2} + 1\right]}$$

Substituting $j\omega$ into the aforementioned equation we get the frequency characteristic:

$$G(j\omega) = \frac{7.5\left(\frac{j\omega}{3}+1\right)}{(j\omega)\left(\frac{j\omega}{2}+1\right)\left[\left(\frac{j\omega}{\sqrt{2}}\right)^2 + \frac{j\omega}{2} + 1\right]}$$

From the above equation, one can get the phase frequency characteristic:

$$\varphi(\omega) = -90^\circ - \arctan\frac{\omega}{2} - \arctan\frac{\omega}{2-\omega^2} + \arctan\frac{\omega}{3}$$

From the aforementioned analysis, for the proportion link, $K = 7.5$, $20\lg K = 17.5$ dB. The break frequencies from small to big are 1.414 (second-order oscillation link), 2 (inertial link), and 3 (first-order differential link). The scale for X coordinate is related to the break frequency. If the value of break frequency of every typical link is closed to each other, we can choose a bigger distance for every decade. The logarithm amplitude characteristic diagram is shown in Fig. 7.28. The exact points, which are the intersections of the asymptote and the Y coordinate or the break frequency coordinate, should be calculated for plotting the logarithm amplitude characteristic diagram. Here, we will consider one point as an example for calculating the point value.

Using eq. (7.32), the intersection B of low frequency and the break frequency coordinate is

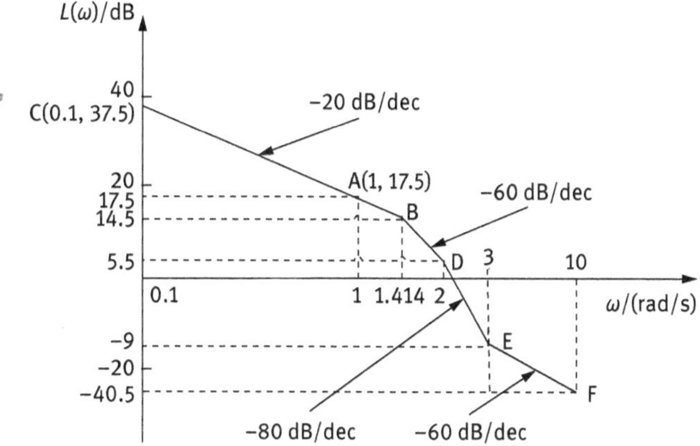

Fig. 7.28: The logarithm amplitude characteristic diagram for Example 7.6.

$$L(\omega) = 20\lg\frac{K}{\omega^N} = 20\lg\frac{7.5}{1.414} = 14.5$$

or

$$\frac{17.5 - L(\omega)}{\lg 1 - \lg 1.414} = -20$$

We can get other points, such as C, D, E, F, by using eq. (7.32) or two-point formula. Note that for low-frequency asymptote, we can use eq. (7.32) to get the intersection, such as points A, B, and C. However, for high-frequency asymptote, one must use the two-point formula rather than eq. (7.32) to get the intersection, such as points D, E, and F.

According to the function, the phase characteristic curves for every typical link should be plot and superimpose them in the same frequency to get the final logarithmic phase characteristic curve. Table 7.2 lists every $\varphi(\omega)$ in every ω.

Table 7.2: Every $\varphi(\omega)$ in every ω for Example 7.6.

ω	0	0.5	1	1.5	2	4	8
$\varphi(\omega)/(°)$	-90	–110.5	–143.1	–180.8	–241.6	–264.4	–268.7

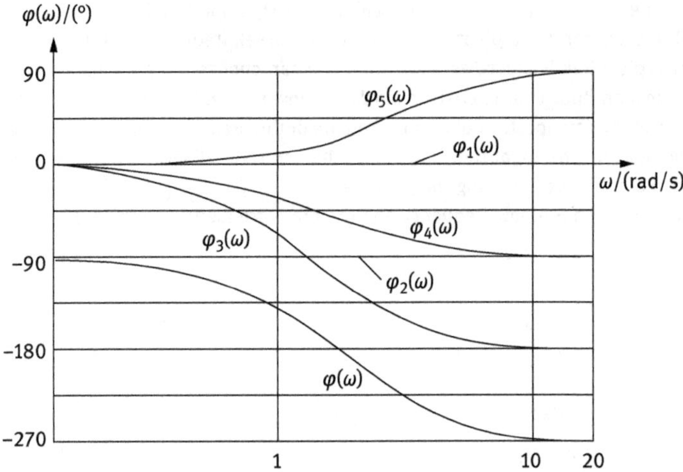

Fig. 7.29: The logarithm phase characteristic diagram for Example 7.6.

In Fig. 7.29, $\varphi_1(\omega)$ is the logarithm phase frequency characteristic of the proportion link; $\varphi_2(\omega)$ is for the integral link; $\varphi_3(\omega)$ is for the second-order oscillation link; $\varphi_4(\omega)$ is for the inertial link; $\varphi_5(\omega)$ is for the first-order differential link; and $\varphi(\omega)$ is the final logarithm phase characteristic diagram for the whole system. We could get it by means of the points depiction method or superposition method according to $\varphi_1(\omega)$, $\varphi_2(\omega)$, $\varphi_3(\omega)$, $\varphi_4(\omega)$, and $\varphi_5(\omega)$.

7.4 Minimum Phase Systems

In the previous examples, the poles and zeros of TF have been restricted to the left-hand side (LHS) of s-plane. However, a system may have zeros and poles located in the right-hand side (RHS) of s-plane, and the range of its phase angle is different depending on how all zeros and poles are located in the LHS of the s-plane. We will use an example to illustrate the new terminology, minimum phase system.

Example 7.7
Consider two systems with the TF given by

$$G_1(s) = \frac{10 \times (0.2s + 1)}{0.1s + 1}$$

$$G_2(s) = \frac{10 \times (0.2s - 1)}{0.1s + 1}$$

Solution
Put $s = j\omega$, and determine the amplitude characteristic function and the phase characteristic function as follows:

$$|G_1(j\omega)| = |G_2(j\omega)| = \frac{10\sqrt{1 + (0.2\omega)^2}}{\sqrt{1 + (0.1\omega)^2}}$$

and

$$\varphi_1(\omega) = \arctan(0.2\omega) - \arctan(0.1\omega)$$

$$\varphi_2(\omega) = -\arctan(0.2\omega) - \arctan(0.1\omega)$$

It is obvious that

$$|G_1(j\omega)| = |G_2(j\omega)|$$

$$|\varphi_1(\omega)| < |\varphi_2(\omega)|$$

On the basis of the above analysis, the phase characteristic of $G_1(s)$, $\varphi_1(\omega)$, has the same change trend with the amplitude characteristic of $G_1(s)$, $|G_1(j\omega)|$, while $G_2(s)$ does not have the same change trend, as shown in Fig. 7.30.

The zeros of a TF are \all reflected about the $j\omega$ axis; there is no change in the gain of the TF, and the unique difference is in the phase-shift characteristics. Actually, the phase change is least for the system with all zeros and poles in the RHS of s-plane.

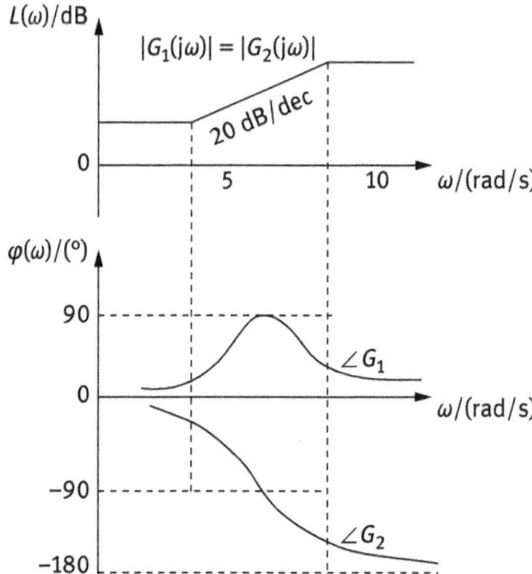

Fig. 7.30: The Bode diagram for Example 7.7.

A system is called a minimum phase system if all its zeros and poles lie in the LHS of the *s*-plane. On the contrary, it is called the nonminimum phase system if it has zeros and (or) poles in the RHS of the *s*-plane.

We consider a system with TF given by

$$G(s) = \frac{K \prod_{k=1}^{p} (T_k s + 1) \prod_{l=1}^{q} (T_l^2 s2 + 2\xi_l T_l s + 1)}{sv \prod_{i=1}^{g} (T_i s + 1) \prod_{j=1}^{h} (T_j^2 s2 + 2\xi_j T_j s + 1)}$$

where $m = p + 2q$, $n = v + g + 2h$, $n \geq$ m. For a minimum phase system, when ω tends ∞ (high frequency), the slope of its logarithm amplitude characteristic is $-20(n - m)$ dB/dec. Thus, for this minimum phase system, when ω tends ∞ (high frequency), its logarithm phase characteristic $\varphi(\infty) = -90°(n - m)$.

For a minimum phase system, the logarithm amplitude characteristic and the logarithm phase characteristic have the same the change trend. In other words, when the slope of the amplitude characteristic increases or decreases, the degree of the phase characteristic also increases or decreases. Therefore, using only its logarithm amplitude characteristic one can confirm the TF of the minimum phase system.

Example 7.8

The logarithm amplitude characteristic asymptote of a minimum phase system is shown in Fig. 7.31. Find its TF.

Solution

To easily analyze, we will give a number to every link, as shown in Fig. 7.32.

Thus, this minimum phase system is composed of a proportion link ①, a first-order differential link ④, and two inertial links ② and ③. Let its TF be

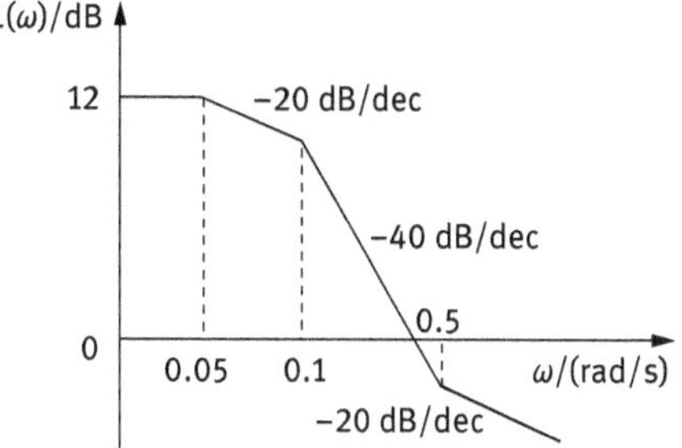

Fig. 7.31: The logarithm amplitude characteristic asymptote for Example 7.8.

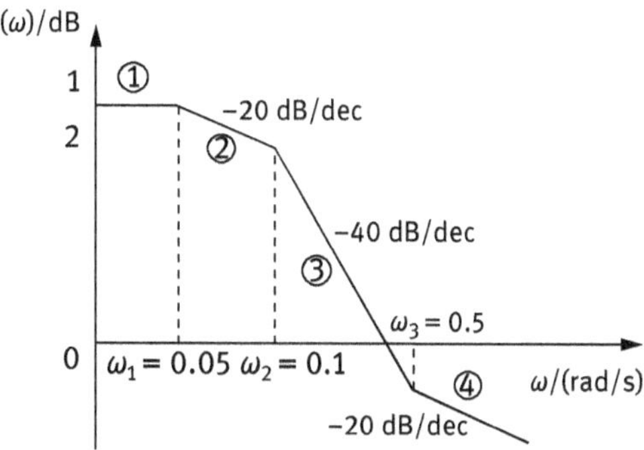

Fig. 7.32: Numbering for every link.

$$G(s) = \frac{K(T_3 s + 1)}{(T_1 s + 1)(T_2 s + 1)}$$

For the proportion link ①, according to $20\lg K = 12$, we can get $K = 4$. And for the inertial link ②,

$$T_1 = \frac{1}{0.05} = 20$$

For the inertial link ③,

$$T_2 = \frac{1}{0.1} = 10$$

For the first-order differential link ④,

$$T_3 = \frac{1}{0.5} = 2$$

Finally, we can get the TF for this minimum phase system,

$$G(s) = \frac{4(2s + 1)}{(20s + 1)(10s + 1)}$$

For this system, we directly give a TF at the beginning according to the minimum phase system. Because it is a minimum phase system, there are no zeros or poles located in the RHS of the s-plane. Thus, the TF of this minimum system can be given as

$$G(s)=\frac{K(T_3s+1)}{(T_1s+1)(T_2s+1)}$$

not

$$G(s)=\frac{K(T_3s-1)}{(T_1s-1)(T_2s+1)}$$

or

$$G(s)=\frac{K(T_3s+1)}{(T_1s-1)(T_2s-1)}$$

and so on.

7.5 Nyquist Stability Criterion

7.5.1 An Explanation for the Nyquist Stability Criterion

The Nyquist stability criterion is a type of geometric criterion and a graphical method in frequency domain, which uses open-loop Nyquist diagram to judge the stability of the closed-loop system. Instead of solving the characteristic roots of the closed-loop system, the criterion gets the stability of the closed-loop system by means of an open-loop frequency characteristic diagram $G(\mathrm{j}\omega)H(\mathrm{j}\omega)$. It is of great significance to use the criterion to analyze the stability of systems, where TF cannot be obtained by mathematical method but frequency characteristic can be obtained by experimental methods. The criterion can not only judge the stability of the closed-loop system but can also judge the stability margin, that is, the relative stability.

The closed-loop system is shown in Fig. 7.33, and its TF is

$$\frac{Y(s)}{X(s)}=\frac{G(s)}{1+G(s)H(s)}$$

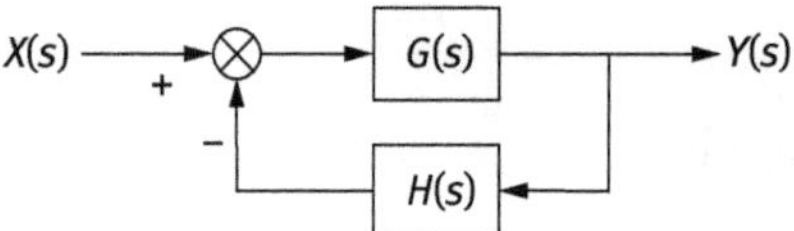

Fig. 7.33: The TF block diagram for the closed-loop system.

The Nyquist stability criterion states that

(1) In the open-loop frequency characteristic $G(j\omega)H(j\omega)$, the closed Nyquist diagram surrounds the point (–1, j0) in anticlockwise direction when ω changes in the range of $-\infty$ to $+\infty$. The number of surrounding is N.

(2) The number of the open-loop poles in the RHS of the s-plane is P. If $N = P$, the closed-loop system is stable.

(3) Since the Nyquist diagram is symmetrical, the closed-loop system is also stable if $2N = P$ when ω changes in the range of 0 to $+\infty$ or $-\infty$ to 0.

Note that for N, anticlockwise is positive and clockwise is negative.

Example 7.9

The open-loop Nyquist diagram is shown in Fig. 7.34. The open-loop TF is

$$G(s)H(s) = \frac{15s^2 + 9s + 1}{(s-1)(2s-1)(3s+1)}$$

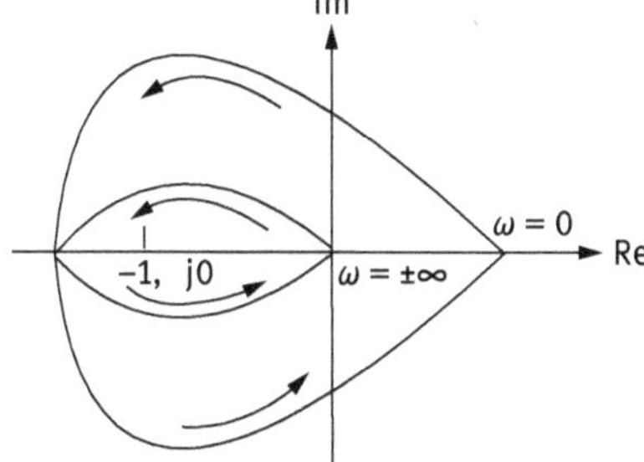

Fig. 7.34: The Nyquist diagram for Example 7.9.

Consider the closed-loop stability of this system.

Solution

According to the denominator of open-loop TF $G(s)H(s)$, there are two poles in the RHS of s-plane, that, is $P = 2$. The closed Nyquist diagram surrounds the point (–1, j0) in anticlockwise direction when the ω changes in the range of $-\infty$ to $+\infty$. The number of surrounding is N. Hence, $N = +2$. Thus, $N = P$, the closed-loop system is stable.

Example 7.10

The open-loop Nyquist diagrams for 4 unit negative feedback systems are shown in Fig. 7.35. In addition, the number of the open-loop unstable characteristic roots, P, is given. Consider closed-loop stability for every system. (The Nyquist diagram is symmetrical.)

Solution

(1) $N = 0$, $P = 0$, $P = 2N$, so the closed-loop system is stable.

(2) $N = 0$, $P = 0$, $P = 2N$, so the closed-loop system is stable.

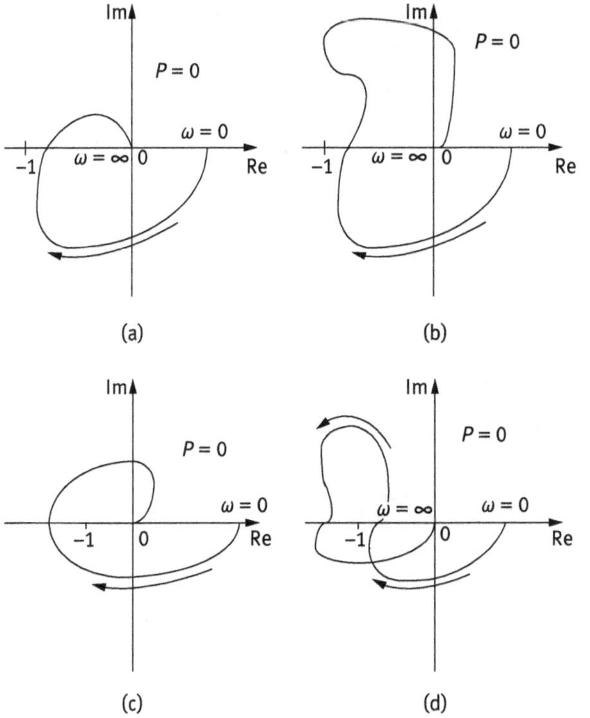

Fig. 7.35: The Nyquist diagrams for Example 7.10.

(3) $P = 0$, when ω changes from 0 to $+\infty$ in Fig. 7.35(c), $N = -1$. When ω changes from $-\infty$ to $+\infty$, $P \neq 2N$, so the closed-loop system is unstable.

(4) $P = 2$, when ω changes from 0 to $+\infty$ in Fig. 7.35(d), $N = 1$. When ω changes from $-\infty$ to $+\infty$, $P = 2N$, so the closed-loop system is stable.

7.5.2 Some Tips for the Nyquist Stability Criterion

To correctly use the open-loop Nyquist diagram to adjust the stability of the closed-loop system, some useful tips are as follows.

(1) When we consider the number of open-loop TF poles in the RHS of s-plane, P, we should also consider open-loop TF poles in the imaginary axis as the LHS poles.

(2) The terminology "crossing" means "the Nyquist diagram crosses the real axis on the left part of the point (–1, j0)."

(3) The number that the open-loop Nyquist diagram surrounds the point (–1, j0) is N.

   ① Once positive crossing N_+ (phase increasing): The Nyquist diagram once crosses the real axis on the LHS of the point (–1, j0) from top to bottom, as shown in Fig. 7.36(a).

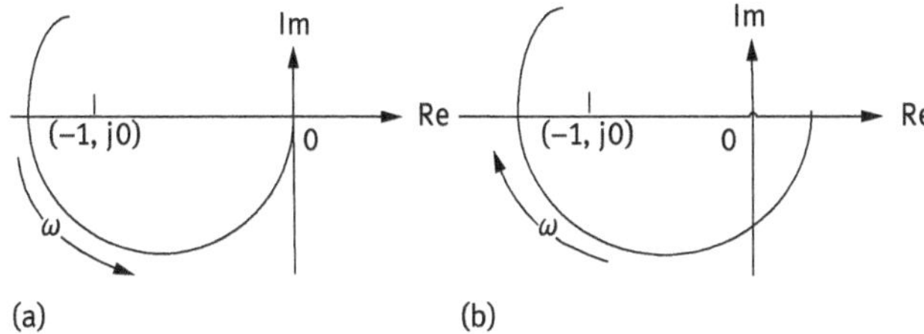

Fig. 7.36: Once crossing: (a) once positive crossing N_+ and (b) once negative crossing N_-.

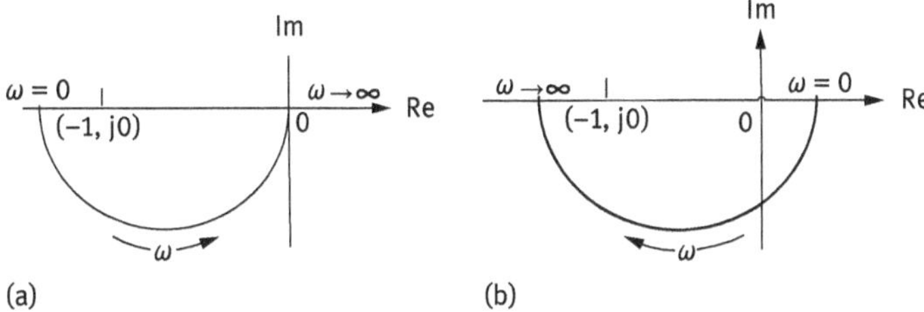

Fig. 7.37: Half crossing: (a) half positive crossing $N_{+1/2}$ and (b) half negative crossing $N_{-1/2}$.

② Once negative crossing N_- (phase decreasing): The Nyquist diagram once crosses the real axis on the LHS of the point (−1, j0) from bottom to top, as shown in Fig. 7.36(b).

③ Half positive crossing $N_{+1/2}$: The Nyquist diagram starts at the real axis on the LHS of the point (−1, j0) from top to bottom, $N_{+1/2} = +1/2$, as shown in Fig. 7.37 (a).

④ Half negative crossing $N_{-1/2}$: The Nyquist diagram ends at the real axis on the LHS of the point (−1, j0) from bottom to top, $N_{-1/2} = -1/2$, as shown in Fig. 7.37(b).

7.5.3 The Nyquist Stability Criterion for a Minimum Phase System

For a minimum phase system, the number of open-loop poles in the RHS of s-plane, P, is always equal to 0 (which means the open-loop system is stable) because there are no open-loop poles in the RHS of s-plane. The required condition for the closed-loop system to be stable for the minimum phase system is $N = 0$.

In Fig. 7.38(a), the required condition for the closed-loop system to be stable for a minimum phase system is that the open-loop Nyquist diagram should not surround point (−1, j0). In Fig. 7.38(b), the open-loop Nyquist diagram crosses the point (−1, j0), and so the closed-loop system is the critical state. In Fig. 7.38(c), the open-loop Nyquist diagram surrounds the point (−1, j0), and so the closed-loop system is unstable.

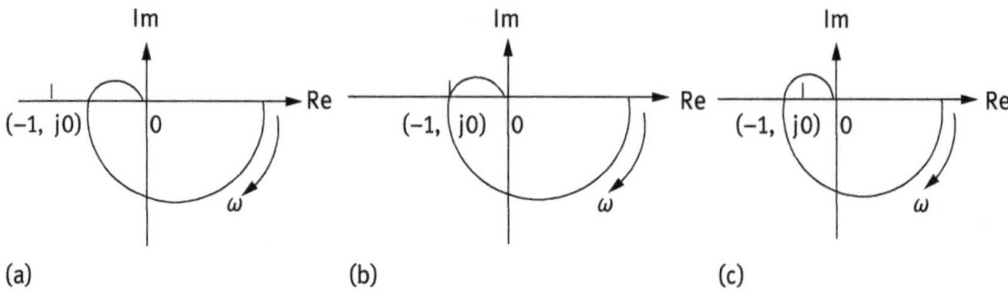

Fig. 7.38: An open-loop Nyquist diagram for the minimum phase system: (a) stable system; (b) critical stable system; and (c) unstable system.

7.5.4 Stability Margin

In linear control system, we can use the Routh –Hurwitz stability criterion, which will be introduced in detail in Chapter 8, to know whether the system is stable or not. However, the Routh –Hurwitz stability criterion cannot determine how stable the system is. Thus, in this case, we can use the Nyquist criterion to judge the degree of stability. From the Nyquist criterion, we could infer that a system is stable with its open-loop and closed-loop form:

(1) More far its open-loop Nyquist diagram from point (−1, j0), more stable is its closed-loop system.
(2) On the contrary, the nearer its open-loop Nyquist diagram is from the point (−1, j0), the lower is its closed-loop system's stability.

7.5.4.1 Introduction to Stability Margin

In Fig. 7.39(a), when the open-loop Nyquist diagram surrounds the point (−1, j0), $N \neq 0$, but $P = 0$. Thus, the closed loop is unstable. The unit step response diagram is not convergent, as shown in the bottom of Fig. 7.39(a). In Fig. 7.39(b), when the open-loop Nyquist diagram crosses the point (−1, j0), the system is in a critical state. The unit step response diagram of its closed-loop system is an oscillation curve with constant amplitude, shown in the bottom of Fig. 7.39(b). In Fig. 7.39(c) or (d), when the open-loop Nyquist diagram does not surround the point (−1, j0), $N = 0$. Hence, the closed-loop system is stable. The unit step response diagram of its closed-loop system is convergent, as shown in the bottom of Fig. 7.39(c) or (d).

From Fig. 7.39(c) and (d), the distance between the open-loop Nyquist diagram and the point (−1, j0) represents the stable degree of the closed-loop system. The system in Fig. 7.39(d) is more stable than that in Fig. 7.39(c) because the distance in Fig. 7.39(d) is larger than that of Fig. 7.39(c). This is the relative stability, also called the stability margin.

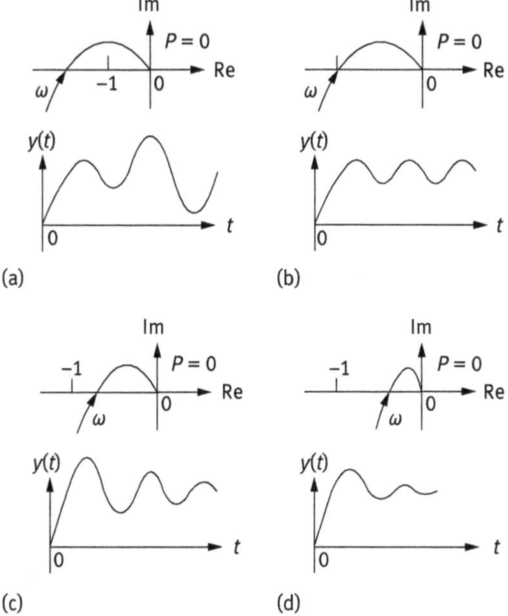

Fig. 7.39: Location cases about the open-loop Nyquist diagram and point (−1, j0): (a) surrounding the point (−1, j0); (b) crossing the point (−1, j0); (c) and (d) not surrounding (−1, j0).

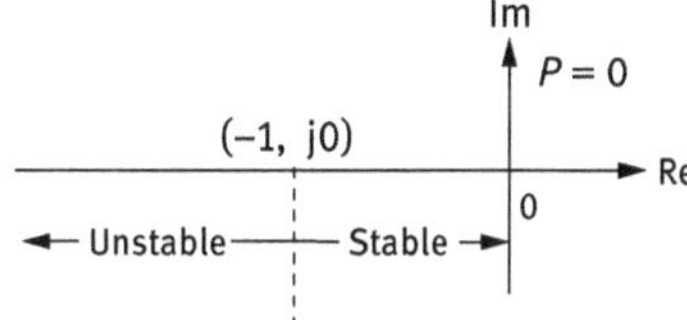

Fig. 7.40: The effect of the point (−1, j0).

From Fig. 7.40, when open-loop Nyquist diagram crosses the real axis on the LHS of the point (−1, j0), $N \neq 0$, the closed-loop system is unstable. The farther its open-loop Nyquist diagram is from point (−1, j0), the more unstable is its closed-loop system. When open-loop Nyquist diagram crosses the real axis on the RHS of the point (−1, j0), $N = 0$, the closed-loop system is stable. The farther its open-loop Nyquist diagram is from point (−1, j0), the more stable is its closed-loop system.

7.5.4.2 The Nyquist Diagram Representation for Stability Margin

From Fig. 7.41, the relationships between the Nyquist diagram and Bode diagram are as follows:

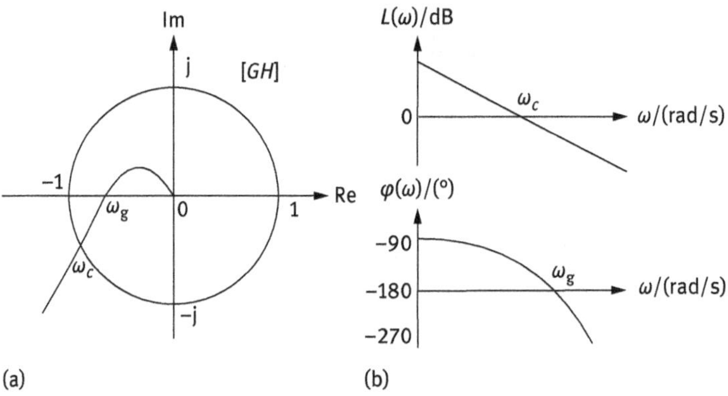

Fig. 7.41: The corresponding relationships between (a) the Nyquist diagram and (b) the Bode diagram.

(1) The unit circle in the Nyquist diagram is equivalent to the ω axis in the Bode amplitude characteristic diagram.

For the Nyquist diagram, the amplitude of the unit circle is 1, and so the logarithm amplitude frequency characteristic should be $20\lg|G(j\omega)H(j\omega)| = 20\lg 1 = 0$ dB. Thus, the line of 0 dB should be the X coordinate for the Bode diagram. Hence, the Nyquist diagram of some control system intersects with the unit circle at the frequency ω_c.

(2) The negative real axis in the Nyquist diagram is equivalent to the horizontal line of −180° in the Bode phase characteristic diagram.

The negative real axis in the Nyquist diagram represents −180°. In the Bode phase characteristic diagram, the line of −180° is parallel with the X coordinate. Thus, the Nyquist diagram of some control system intersects with the −180° or the negative axis at the frequency ω_g.

We can use the phase margin γ and amplitude margin K_g to quantitatively represent the relative stability of the system, as shown in Fig. 7.42.

(1) **Phase Margin** γ

From Fig. 7.42(a) and (b), the unit circle intersects with the Nyquist diagram at point A. Connecting the original point O and A we get the straight line OA. The angle between the straight line OA and the negative real axis is called the phase margin γ,

$$\gamma = \varphi(\omega_c) - (-180^\circ) = \varphi(\omega_c) + 180^\circ \tag{7.57}$$

where ω_c is the amplitude crossing frequency.

(2) **Amplitude Margin K_g**

From Fig. 7.42 (a) and (b), the open Nyquist diagram intersects with the negative real axis at point Q and the amplitude of the point Q is $|G(jw_g)H$

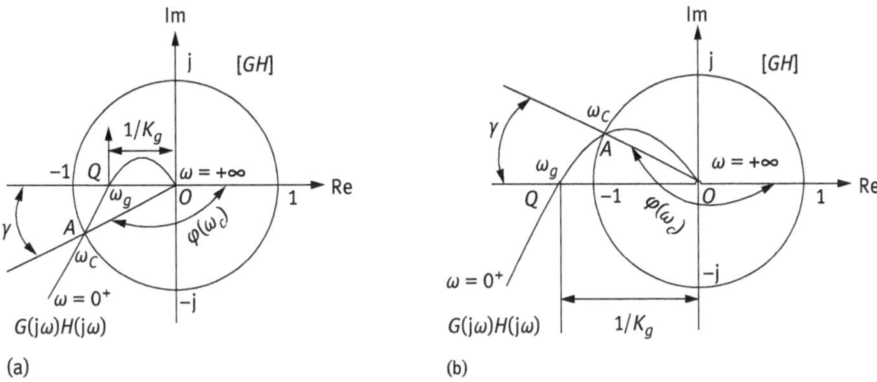

Fig. 7.42: The phase margin γ and amplitude margin K_g: (a) positive phase margin and positive amplitude margin and (b) negative phase margin and negative amplitude margin.

$(j\omega_g)|$. Introducing a new terminology K_g and letting K_g represent the reciprocal of $|G(jw_g)H(jw_g)|$,

$$K_g = \frac{1}{|G(j\omega_g)H(j\omega_g)|} \tag{7.58}$$

ω_g is the phase crossing frequency. K_g is known as the amplitude margin . The decibel value of the amplitude margin K_g is K_g(dB), and K_g(dB) is common for evaluating the degree of stability,

$$K_g(dB) = 20\lg K_g = 20\lg\left|\frac{1}{G(j\omega_g)H(j\omega_g)}\right| = -20\lg|G(j\omega_g)H(j\omega_g)| \tag{7.59}$$

For the closed-loop systems (its open-loop system is stable), if $\gamma > 0$, K_g (dB) > 0 (or $1/K_g < 1$ or $K_g > 1$), it is stable, as shown in Fig. 7.42(a). If $\gamma < 0$, K_g (dB) < 0(or $1/K_g > 1$ or $K_g < 1$), it is unstable, as shown in Fig. 7.42(b). However, overstable margin may effect the other performance of the system, such as the response rapidness of the control system. Hence, we usually select some range of stable margin in engineering practical process, K_g (dB) = (6 ~ 20) dB, $\gamma = 30° \sim 60°$.

7.5.4.3 The Bode diagram representation for a stable margin

(1) **Stable System**

For the stable system, the phase margin and the amplitude margin are positive. Hence, in the Bode diagram, the angle g is beyond the straight line of −180° (γ >0), and K_g (dB) is below the ω axis ($K_g > 1$, K_g (dB) > 0), as shown in Fig. 7.43(a).

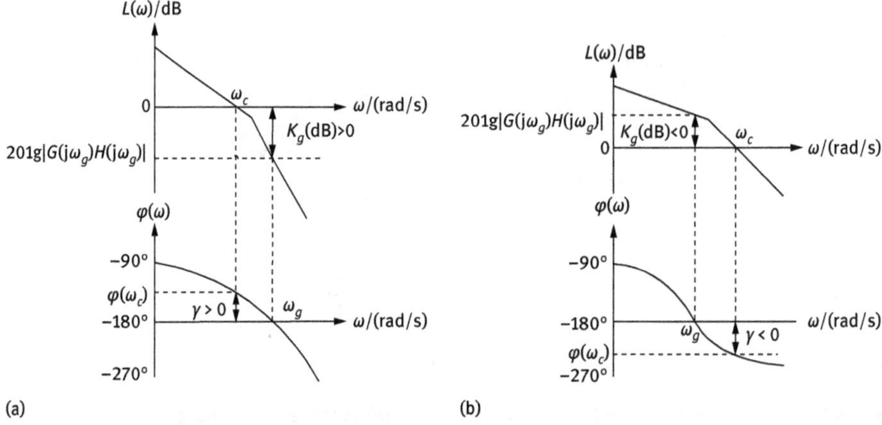

Fig. 7.43: The Bode diagram representations for stability margin: (a) the stable system and (b) the unstable system.

(2) **Unstable system**
For the unstable system, the phase margin and the amplitude margin are negative. In the Bode diagram, the angle γ is below the straight line of −180° ($\gamma < 0$), and K_g (dB) is beyond the ω axis ($K_g < 1$, K_g (dB) < 0), as shown in Fig. 7.43(b).

Example 7.11
The open-loop TF of some system is

$$G(s)H(s) = \frac{20}{s(0.2s+1)}$$

Evaluate the amplitude margin K_g (dB) and the phase margin γ.

Solution
The open-loop frequency characteristic is

$$G(j\omega)H(j\omega) = \frac{20}{j\omega(1+0.2j\omega)} \tag{7.60}$$

The amplitude frequency characteristic is

$$|G(j\omega)H(j\omega)| = \frac{20}{\omega\sqrt{1+0.04\omega^2}} \tag{7.61}$$

The phase frequency characteristic is

$$\angle G(j\omega)H(j\omega) = -90^\circ - \arctan 0.2\omega \tag{7.62}$$

Thus, from eqs. (7.59) and (7.57), we have

$$K_g = -20\lg|G(j\omega_g)H(j\omega_g)| = -20\lg\frac{K}{\omega_g\sqrt{1+0.04\omega_g^2}} \tag{7.63a}$$

$$\gamma = \varphi(\omega_c) + 180^\circ = -90^\circ - \arctan 0.2\omega_c + 180^\circ \tag{7.63b}$$

When $\omega = \omega_g$, the phase frequency characteristic should be a line of negative 180°. That is,

$$\angle G(j\omega_g)H(j\omega_g) = -180^\circ$$

Hence, from eq. (7.62), we have

$$\angle G(j\omega_g)H(j\omega_g) = -90^\circ - \arctan 0.2\omega_g = -180^\circ \tag{7.64}$$

Solving it we get the value of ω_g,

$$\omega_g = \infty$$

And when $\omega = \omega_c$, the amplitude frequency characteristic should be equal unit,

$$|G(j\omega_c)H(j\omega_c)| = 1$$

So, from eq. (7.61), we have

$$|G(j\omega_c)H(j\omega_c)| = \frac{K}{\omega_c\sqrt{1+0.04\omega_c^2}} = 1 \tag{7.65}$$

Furthermore, we have

$$0.04\omega_c^4 + \omega_c^2 - 400 = 0$$

Solving it we get the value of ω_c,

$$\omega_c = 9.35$$

Finally, substituting ω_g and ω_c in eqs. (7.63a) and (7.63b) we get the exact value for amplitude margin K_g (dB) and the phase margin γ,

$$\begin{cases} K_g = -20\lg|G(j\omega_g)H(j\omega_g)| = -20\lg\frac{20}{\omega_g\sqrt{1+0.04\omega_g^2}} = \infty \text{ dB} \\ \gamma = \varphi(\omega_c) + 180^\circ = -90^\circ - \arctan 0.2\omega_c + 180^\circ = 28.14^\circ \end{cases}$$

7.6 Problems

P7.1. The TF of the system is

$$\frac{Y(s)}{X(s)} = \frac{10}{0.5s+1}$$

The input signal is $x(t) = X\sin(\omega t)$ and the frequency f and the amplitude X of the input sine signal are 1 Hz and 10, respectively. Find the frequency response output $y(t)$. (Hint: $\omega = 2\pi f$.)

P7.2. Plot the Bode diagram for the system with the following TF:

$$G(s) = \frac{1250(s+2)}{s^2(s^2+6s+25)}$$

P7.3. The open-loop TF of some system is

$$G(s)H(s) = \frac{20}{s(0.5s+1)}$$

Evaluate the amplitude margin K_g (dB) and the phase margin γ.

P7.4. The logarithmic amplitude characteristic asymptotes of minimum phase systems are shown in Fig. 7.44. Find their TFs for every minimum phase system.

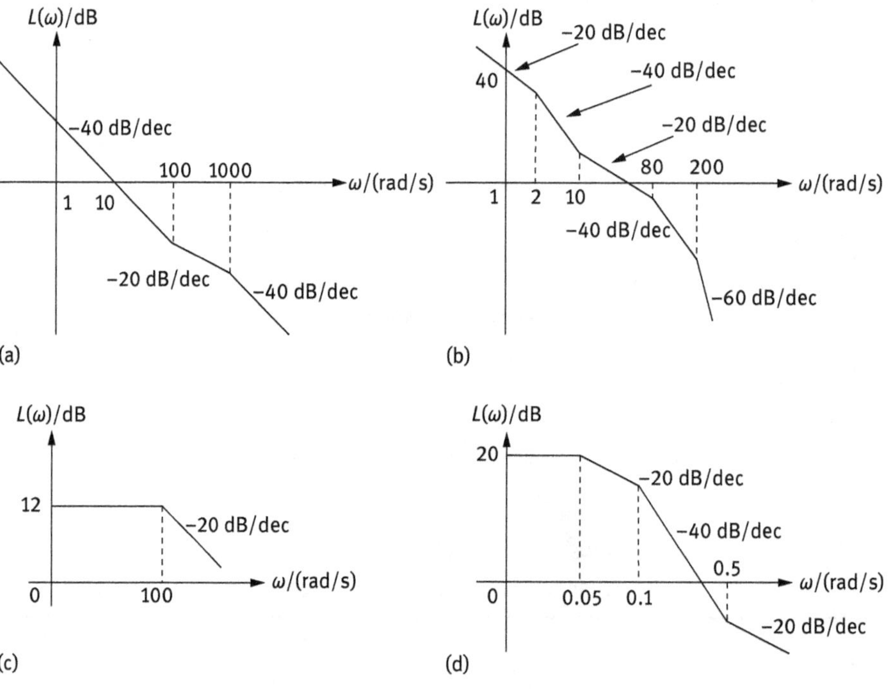

Fig. 7.44: The logarithmic amplitude characteristic asymptotes for P7.4.

8 Stability Analysis of Control Systems

Stability analysis of a control system is essential for linear control system construction. The chapter introduces the concept of stability in Section 8.1. The condition for the system stability is discussed in Section 8.2 together with the necessary conditions for the stability of the linear feedback system. In Section 8.3, we will focus on the Routh–Hurwitz stability criterion, especially for some special cases for the stability criterion.

8.1 Stability

When considering analysis and design of a control system, stability analysis is always presented with paramount significance. Generally, a system is said to be stable if its response is bounded. On the contrary, if the response of the system fluctuates or even deviates without proper time constraints, then the system is regarded as unstable.

In Section 8.2, we show that if the closed-loop poles lie on the LHS of the s-plane, the control system is stable, as shown in Fig. 8.1. Hence, a feedback system is necessarily regarded as stable when all poles of the closed-loop transfer function (TF) have negative real parts.

As Fig. 8.2(a) illustrates, the two curves represent the stable systems, especially curve no. 1 for damping oscillation, and curve no. 2 for single peak convergence. While for Fig. 8.2(b), the two curves represent the unstable systems, where curve no. 1 is for the constant oscillation, and curve no. 2 for the divergent oscillation.

On the basis of the above analysis, if the system is stable, its stability is reflected as follows:

(1) The output of the system can move to a new steady state when it is subjected to an input signal.

(2) The output of the system can recover to the initial stable value after the disturbance input signal is removed.

8.2 Conditions for the System Stability

From Fig. 8.3, TF of this control system is

$$\frac{Y(s)}{X(s)} = \frac{G_1(s)G_2(s)}{1 + G_1(s)G_2(s)H(s)} = \frac{b_m s^m + b_{m-1}s^{m-1} + \cdots + b_1 s + b_0}{a_n s^n + a_{n-1}s^{n-1} + \cdots + a_1 s + a_0} \tag{8.1}$$

The eigen equation and homogeneous differential equation of the above TF and block diagram are

https://doi.org/10.1515/9783110573275-008

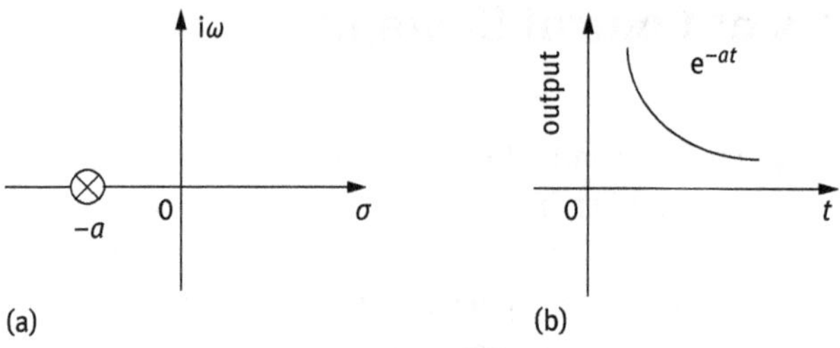

Fig. 8.1: A stable system: (a) the closed-loop pole and (b) the response curve.

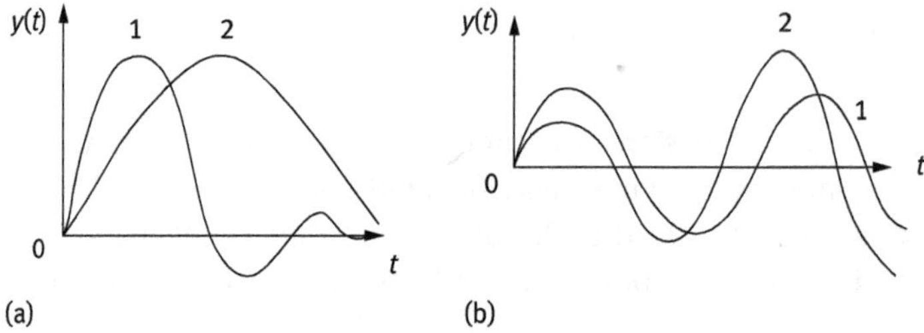

Fig. 8.2: The oscillation styles for different systems: (a) constant amplitude oscillation and (b) divergent oscillation.

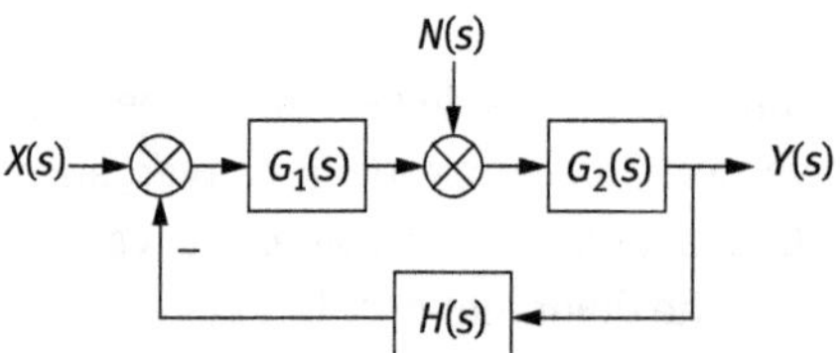

Fig. 8.3: A function block diagram.

$$a_n s^n + a_{n-1} s^{n-1} + \cdots + a_1 s + a_0 = 0 \tag{8.2}$$

$$a_n \frac{\mathrm{d}^n y}{\mathrm{d}t^n} + a_{n-1} \frac{\mathrm{d}^{n-1} y}{\mathrm{d}t^{n-1}} + \cdots + a_1 \frac{\mathrm{d}y}{\mathrm{d}t} + a_0 y = 0 \tag{8.3}$$

The general solution of eq. (8.3) is

$$y(t) = c_1 \mathrm{e}^{s_1 t} + c_2 \mathrm{e}^{s_2 t} + \cdots + c_n \mathrm{e}^{s_n t} \tag{8.4}$$

The solution of eq. (8.4) is

$$y(t) = \sum_{i=1}^{k} D_i \mathrm{e}^{s_i t} + \sum_{i=1}^{r} \mathrm{e}^{\delta_j t} (E_i \cos\omega_i t + F_i \sin\omega_i t) \tag{8.5}$$

where the number of the real roots is k, and the number of the complex is $2r$. When $\mathbf{s}_i$ is less than 0, δ_j is less than 0 as t approches to infinity. The value of $y(t)$ converges to zero. Taking the real number poles and the complex poles into consideration, eq. (8.2) can be factorized as

$$\prod_{i=1}^{k} a_n(s-s_i)\prod_{i=1}^{r}[s-(\delta_i+j\omega_i)]\cdot[s-(\delta_i-j\omega_i)]=0 \tag{8.6}$$

By means of the aforementioned analysis, we can reach the relevant conclusion about the sufficient and necessary conditions for the stability of linear feedback system:

1. All poles of the system must be in the LHS of s-plane (imaginary axis excluded).
2. For the closed-loop TF, the characteristic roots must cosatisfy the following four conditions:
   ① no nil solution;
   ② no conjugate pure complex roots, that is, $\mathrm{Re}(s) = 0$ (since the conjugate pure complex roots imply that the output of the system is a constant amplitude oscillation curve);
   ③ all the real roots must be negative; and
   ④ all the real parts of conjugate complex roots must be negative.

As a matter of fact, the aforementioned two aspects are two kinds of expressions with the same meaning.

Taking the open-loop TF of the unit negative feedback system as an example,

$$G(s)=\frac{k}{s(Ts+1)}$$

where $T > 0$, $k > 0$, and $1–4Tk < 0$. The closed-loop TF of the system is

$$\Phi(s)=\frac{G(s)}{1+G(s)}=\frac{k}{Ts^2+s+k}$$

The characteristic equation (CE) of the above closed-loop TF is

$$Ts^2+s+k=0$$

Where its eigen values are

$$s_{1,2}=\frac{-1\pm\sqrt{1-4Tk}}{2T}=-\frac{1}{2T}\pm\frac{\sqrt{4Tk-1}}{2T}j$$

The real part of the above eigen values is negative, that is, the above example possesses negative real parts in its CE, which indicates that the closed-loop system is steady.

8.3 Routh–Hurwitz Stability Criterion

From the foregoing section, it is necessary to identify the poles of the closed-loop TF in order to determine the stability of a feedback control system. Why Routh–Hurwitz stability criterion is applied to investigate stability of feedback systems? Routh–Hurwitz stability criterion is a rigorous approach for determining the stability of the system without actually solving the roots of the CE. In this section, we will introduce this rigorous method with details.

8.3.1 The Preconditions for Routh–Hurwitz Stability Criterion

The CE of the system shown in Fig. 8.3 is eq. (8.2):

$$a_n s^n + a_{n-1} s^{n-1} + \cdots + a_1 s + a_0 = 0$$

The preconditions before applying the Routh–Hurwitz stability criterion are

(1) All coefficients of the CE, including a_n, a_{n-1}, ..., a_1, a_0, are not equal to zero. That is, all coefficients of CE must exist.

(2) The sign of all coefficients must be the same.

The above two conditions are the preconditions; further conditions needs to be supplemented.

8.3.2 Full Condition for Routh–Hurwitz Stability Criterion

For CE (8.2), its Routh array or schedule is

$$\begin{array}{cccccc} s^n & a_n & a_{n-2} & a_{n-4} & a_{n-6} & \cdots \\ s^{n-1} & a_{n-1} & a_{n-3} & a_{n-5} & a_{n-7} & \cdots \\ s^{n-2} & b_1 & b_2 & b_3 & b_4 & \cdots \\ s^{n-3} & c_1 & c_2 & c_3 & c_4 & \cdots \\ \cdots & \cdots & \cdots & \cdots & \cdots & \cdots \\ s^1 & d_1 & & & & \\ s^0 & e_1 & & & & \end{array} \tag{8.7}$$

The first row of eq. (8.7) is the coefficients of all even terms or odd terms of s, with descending power of s. The second row is the coefficients for all odd terms or even terms of s, and is ranked according to descending order of s. Other coefficients are yielded by the above two rows,

$$b_1 = \frac{a_{n-1}a_{n-2} - a_n a_{n-3}}{a_{n-1}} \tag{8.8a}$$

$$b_2 = \frac{a_{n-1}a_{n-4} - a_n a_{n-5}}{a_{n-1}} \tag{8.8b}$$

$$b_3 = \frac{a_{n-1}a_{n-6} - a_n a_{n-7}}{a_{n-1}} \tag{8.8c}$$

$$b_4 = \frac{\cdots}{\cdots} \tag{8.8d}$$

$$c_1 = \frac{b_1 a_{n-3} - a_{n-1} b_2}{b_1} \tag{8.9a}$$

$$c_2 = \frac{b_1 a_{n-5} - a_{n-1} b_3}{b_1} \tag{8.9b}$$

$$c_3 = \frac{b_1 a_{n-7} - a_{n-1} b_4}{b_1} \tag{8.9c}$$

$$c_4 = \frac{\cdots}{\cdots} \tag{8.9d}$$

New rows are generated by previous two rows. Finally, we obtain Routh array by (n+1) rows. There is only one element in the last two rows with other coefficients evaluated to zero.

We name a_n, a_{n-1}, b_1, c_1, ..., d_1, e_1, as the first-column elements of Routh array. The statement for the sufficient and necessary condition of the Routh–Hurwitz stability criterion is as follows:

For a stable system, there should be no sign transformation in the first column of the Routh array. Otherwise, the system is unstable. The number of sign transformations in the first column is equal to the number of the roots that lies on the RHS of S-plane for the closed-loop system.

The Routh–Hurwitz stability criterion is applied to determine the stability of various complex systems. The following examples are presented to illustrate its applications and procedures in stability analysis.

Example 8.1

The closed-loop TF of the control system is

$$\Phi(s) = \frac{6s+4}{s^4 + 7s^3 + 17s^2 + 17s + 6}$$

Identify the stability of the system by using Routh–Hurwitz stability criterion.

Solution

The CE of the closed-loop TF is

$$s^4 + 7s^3 + 17s^2 + 17s + 6 = 0$$

The signs of the coefficients in the eigen equation are all positive with all terms existing. That is, it meets preconditions. According to the Routh array to perform in-depth analysis,

$$\begin{array}{llll} s^4 & 1 & 17 & 6 \\ s^3 & 7 & 17 & 0 \\ s^2 & \frac{7\times17-1\times17}{7} = 14.57 & \frac{7\times6-1\times0}{7} = 6 & 0 \\ s^1 & \frac{14.57\times17-7\times6}{14.57} = 14.12 & 0 & 0 \\ s^0 & \frac{14.12\times6-14.57\times0}{14.12} = 6 & 0 & 0 \end{array}$$

Since there is no alteration in the sign in the first column, no root has positive real part, which implies that the feedback control system mentioned in question stem is stable.

Example 8.2

The open-loop TF of the unit negative feedback control system is

$$G(s) = \frac{K}{s(s+1)(s+2)}$$

Evaluate the range of K in order to ensure that the closed-loop system is stable.

Solution

The closed-loop TF for this system is

$$\Phi(s) = \frac{G(s)}{1+G(s)} = \frac{K}{s^3 + 3s^2 + 2s + K}$$

Hence, the CE is

$$s^3 + 3s^2 + 2s + K = 0$$

First, according to the precondition for Routh–Hurwitz stability criterion, all coefficients must be larger than zero. So $K > 0$. Secondly, applying Routh–Hurwitz stability criterion to determine the stability of the system. The Routh array is

$$\begin{array}{lll} s^3 & 1 & 2 \\ s^2 & 3 & K \\ s^1 & \frac{6-K}{3} & 0 \\ s^0 & K & 0 \end{array}$$

According to the Routh–Hurwitz stability criterion, we have

$$\begin{cases} K > 0 \\ \frac{6-K}{3} > 0 \end{cases}$$

That is,

$$0 < K < 6$$

It is often crucial to assess the stability degree of a stable system. To judge its relative stability, we need to know how close are the roots to the Y-axis. This can be implemented by translating the Y-axis to the left and by finding whether the shift results are unstable in the new axis. The shift of the Y-axis to −a means that all the values of s are replaced by $u - a$ ($s = u - a$ or $u = s + a$) in CE, where the equation with u is now being tested for stability.

In the frequency domain analysis field, the stability margin (or relative stability) is represented by the distance a to the imaginary axis. The distance a is the minimum absolute value of the real part among all characteristic roots. In fact, we have discuss this in the last chapter. The characteristic roots should be on the LHS of the line $s = -a$ for the system with the stability margin a, as shown in the shadow of Fig. 8.4.

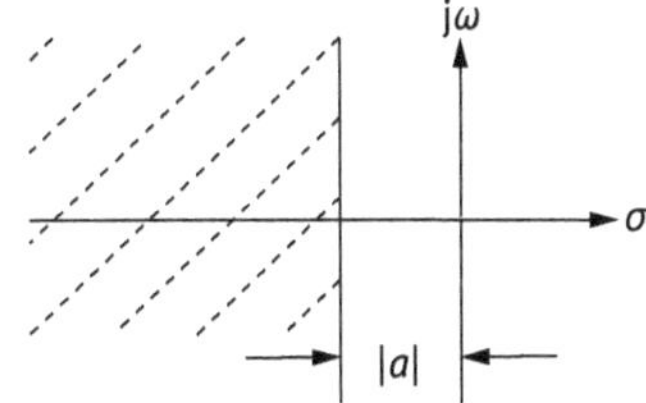

Fig. 8.4: The location of the characteristic roots for systems with stability margin a.

Example 8.3
The open-loop TF of the unit negative feedback control system is

$$G(s) = \frac{K}{s\left(\frac{s}{3}+1\right)\left(\frac{s}{6}+1\right)}$$

Evaluate the range of K if all the real parts of the roots for the closed-loop CE are less than −1 (or evaluate the value range of K in order to keep the stability margin a equal to 1). How about less than −2?

Solution
The closed-loop TF of this unit negative feedback control system is

$$\Phi(s) = \frac{18K}{s^3 + 9s^2 + 18s + 18K}$$

where its CE is

$$s^3 + 9s^2 + 18s + 18K = 0$$

Let s be $u-1$ and substitute the aforementioned equation to obtain a new CE upon variable u,

$$u^3 + 6u^2 + 3u + (18K - 10) = 0$$

According to Routh–Hurwitz stability criterion precondition, $18K - 10 > 0$, that is, $K > 5/9$. Then the Routh array is

u^3	1	3
u^2	6	$18K-10$
u^1	$\frac{14-9K}{3}$	0
u^0	$18K-10$	0

If all the real parts of roots in the closed-loop CE are less than −1, the range of K is

$$\begin{cases} \frac{14-9K}{3} > 0 \\ 18K-10 > 0 \end{cases} \rightarrow \frac{5}{9} < K < \frac{14}{9}$$

Then, let $s = u-2$ and generate another CE about variable u,

$$u^3 + 3u^2 - 6u + (18K - 8) = 0$$

Irrespective of the value of K, all the real parts of the roots for the closed-loop CE are not less than −2 due to the signs of the coefficients for the CE change. In fact, here, the stability margin a for this control system belongs to the range [−2, −1]. In other words, the real parts of CE roots also belong to this range, shown in Fig. 8.5.

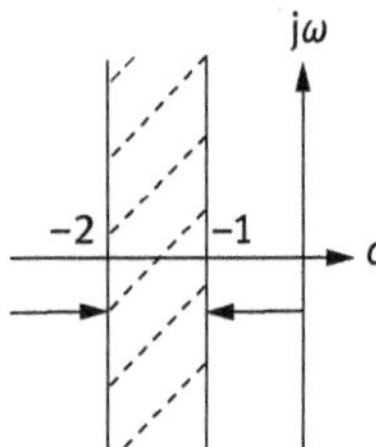

Fig. 8.5: The distribution of the real parts of CE roots for Example 8.3.

Example 8.4

The open-loop TFs of unit negative feedback control systems are as follows:

$$G(s)H(s) = \frac{k}{(0.5s+1)^3},\ G(s)H(s) = \frac{k}{(0.05s^2+1)(0.05s+1)}$$

Analyze the stabilities of these two systems by the Routh–Hurwitz stability criterion when $k = 6$ and $k = 15$, respectively. What conclusions could one draw from the results?

Solution (1)

The closed-loop TF is

$$\Phi(s) = \frac{8k}{s^3 + 6s^2 + 12s + 8(1+k)}$$

The CE is

$$s^3 + 6s^2 + 12s + 8(1+k) = 0$$

The Routh array is

$$\begin{array}{lcc} s^3 & 1 & 12 \\ s^2 & 6 & 8(1+k) \\ s^1 & \frac{32-4k}{3} & 0 \\ s^0 & 8(1+k) & 0 \end{array}$$

Hence,

$$32-4k>0 \text{ and } 1+k>0$$

$$-1<k<8$$

The above analysis implies that the system is stable when $k=6$ but unstable when $k=15$. From the aforementioned solution, we can reach a conclusion that the range of open-loop gain (k) is influential on system stability.

Solution (2)
The closed-loop TF is

$$\Phi(s)=\frac{k}{(0.05)^2s^3+0.05s^2+0.05s+(1+k)}$$

The CE is

$$(0.05)^2s^3+0.05s^2+0.05s+(1+k)=0$$

The Routh array is

$$\begin{array}{lcc} s^3 & 0.05^2 & 0.05 \\ s^2 & 0.05 & 1+k \\ s^1 & -0.05k & 0 \\ s^0 & 1+k & \end{array}$$

Hence,

$$-0.05k>0,\ 1+k>0,\ -1<k<0$$

The system is unstable when $k=6$ and $k=15$ because they do not belong to the value range of (−1, 0).

8.3.3 Special Cases for the Routh–Hurwitz Stability Criterion

8.3.3.1 Case 1: The first-column element in some row is zero but others in the same row are not all zero

If the first-column element in some row is zero, calculation is forced stopped as the first-column element (i.e., zero) cannot be the denominator while calculating the next row. To address this problem, an infinite small positive number ε is introduced to replace zero for further calculation.

Example 8.5
The CE of a closed-loop TF is

$$s^4+2s^3+s^2+2s+1=0$$

Determine the stability of the system according to the Routh–Hurwitz stability criterion.

Solution
According to the CE, the Routh array is

$$\begin{array}{llll} s^4 & 1 & 1 & 1 \\ s^3 & 2 & 2 & 0 \\ s^2 & 0 & 1 & 0 \\ s^1 & & & \\ s^0 & & & \end{array}$$

The first-column element in the third row is zero, and so we introduce an infinitely small positive number ε to take the place of that zero and carry on the calculation:

$$\begin{array}{lccc} s^4 & 1 & 1 & 1 \\ s^3 & 2 & 2 & 0 \\ s^2 & \varepsilon(\varepsilon\to 0) & 1 & 0 \\ s^1 & 2-\frac{2}{\varepsilon} & 0 & 0 \\ s^0 & 1 & 0 & 0 \end{array}$$

Since $\varepsilon \to 0$,

$$2-\frac{2}{\varepsilon}\to -\infty$$

As the signs of the first-column elements varies twice from positive to negative and to positive again, the closed-loop system is unstable with two roots in the RHS of s-plane.

8.3.3.2 Case 2: All elements in some row are zero

This situation reveals that there are some symmetrical characteristic roots in the system.

(1) Two real roots with the same absolute value and the opposite sign. The response of the system is divergent and the system is unstable, as depicted in Fig. 8.6(a).
(2) Two pairs of conjugate complex roots with the same absolute value in the imaginary part and the opposite sign in the real part. The response is divergent and the system is unstable, as shown in Fig. 8.6(b).

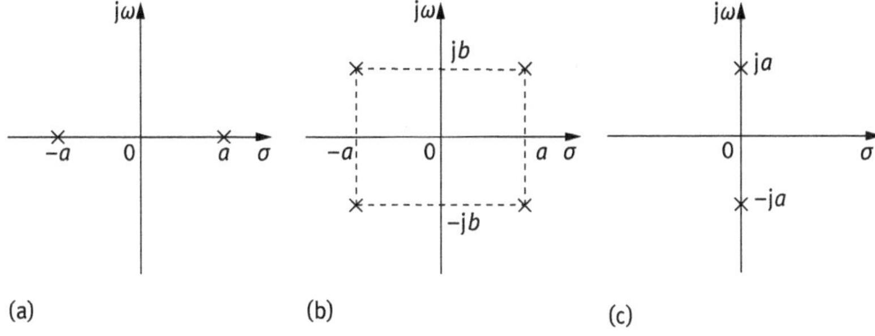

Fig. 8.6: The symmetrical characteristic roots distributions: (a) two real roots with the same absolute value and opposite sign; (b) two pairs of conjugate complex roots symmetrical with the imaginary axis; and (c) a pair of conjugate pure imaginary roots.

(3) A pair of conjugate pure imaginary roots. The response is a constant amplitude oscillation with the special frequency and the system is critically stable, as illustrated in Fig. 8.6(c).

(4) The combination of the above roots.

In this case, the Routh array will be terminated at the row in which all elements are equal to zero. In order to continue calculating the Routh array, one can set an "auxiliary equation" by means of the previous row of the zero-element row. The orders of s in the auxiliary equation are all even and descending. By the derivative of the auxiliary equation we can get a new lower-order differential equation compared to the auxiliary equation and then substitute the zero-element row by the coefficients of the new lower-order equation. Finally, get a new Routh Array. At the same time, the eigen values can be obtained from the auxiliary equation.

Example 8.6

The CE of a closed-loop TF is

$$s^6 + 2s^5 + 8s^4 + 12s^3 + 20s^2 + 16s + 16 = 0$$

Identify stability of the system.

Solution

The Routh array is

$$\begin{array}{lllll} s^6 & 1 & 8 & 20 & 16 \\ s^5 & 2 & 12 & 16 & 0 \\ s^4 & 2 & 12 & 16 & 0 \\ s^3 & 0 & 0 & 0 & 0 \\ s^2 & & & & \\ s^1 & & & & \end{array}$$

The whole elements in the row of the s^3 are zero. Herein one should construct an auxiliary equation with elements in the row over the zero-element row. Characteristic roots can be obtained by this auxiliary equation. Introducing an "auxiliary equation" by the upper row of the zero-element row, we get

$$A(s) = 2s^4 + 12s^2 + 16$$

Taking derivative of the above auxiliary equation, we obtain

$$\frac{\mathrm{d}A(s)}{\mathrm{d}s} = 8s^3 + 24s$$

Substituting the zero-element row by coefficients of the above equation that yields a new Routh array,

$$\begin{array}{ccccc}
s^6 & 1 & 8 & 20 & 16 \\
s^5 & 2 & 12 & 16 & 0 \\
s^4 & 2 & 12 & 16 & 0 \\
s^3 & 8 & 24 & 0 & 0 \\
s^2 & 6 & 16 & 0 & 0 \\
s^1 & \frac{8}{3} & 0 & 0 & 0 \\
s^0 & 16 & 0 & 0 & 0
\end{array}$$

From the first column, there is no change in the signs. Therefore, there are no positive roots or conjugate complex roots with the positive real part. However, the conjugate pure imaginary roots must exist because the coefficients of row s^3 are all zero. Two pairs of conjugate pure imaginary roots can be obtained by solving the auxiliary equation $A(s)$,

$$2s^4 + 12s^2 + 16 = 0$$

Solving this equation, we obtain those two pairs of conjugate pure imaginary roots,

$$\begin{cases} s_{1,2} = \pm\sqrt{2}\mathrm{j} \\ s_{3,4} = \pm 2\mathrm{j} \end{cases}$$

We can conclude that the system is critically stable. As a matter of fact, it is unstable with a constant oscillation in its response.

Example 8.7

A function block diagram is shown in Fig. 8.7. Evaluate the value of K and α, in order that the system oscillates constantly with a frequency of $\omega = 2$ rad/s.

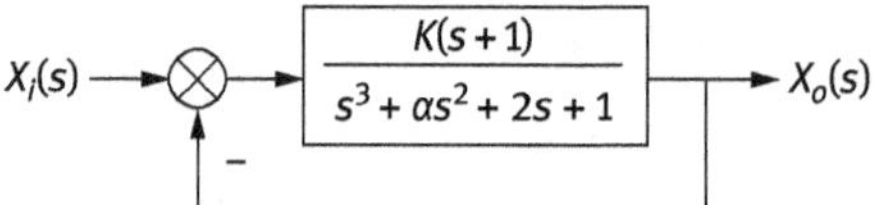

Fig. 8.7: A function block diagram.

Solution

According to the function block diagram, we can derive the closed-loop TF and the CE is $s^3 + \alpha s^2 + (2 + K)s + (1 + K) = 0$. Thus, the Routh array is

$$\begin{array}{lcc} s^3 & 1 & 2+K \\ s^2 & \alpha & 1+K \\ s^1 & (2+K)-\frac{1+K}{\alpha} & 0 \\ s^0 & 1+K & 0 \end{array}$$

Only when the elements in the row of s^1 are all zero can one acquire a pair of conjugate pure imaginary roots instead of two or more pairs, regarding the row of s^2 as the auxiliary equation. The auxiliary equation is

$$\alpha s^2 + (1+K) = 0 \tag{8.10a}$$

The roots of eq. (8.10a) are

$$s_{1,2} = \pm\sqrt{\frac{1+K}{\alpha}}\mathrm{j} \tag{8.10b}$$

Since the system can oscillate constantly with the frequency of $\omega = 2$ rad/s, the conjugate pure imaginary roots are

$$s_{1,2} = \pm 2\mathrm{j} \tag{8.10c}$$

Considering eqs. (8.10b) and (8.10c), we get

$$\sqrt{\frac{1+K}{\alpha}} = 2 \tag{8.10d}$$

From the elements in the row of s^1, we can gain another equation:

$$(2+K) - \frac{1+K}{\alpha} = 0 \tag{8.10e}$$

Finally, from eqs. (8.10d) and (8.10e), the values of K and α can be obtained as follows:

$$K = 2, \alpha = 0.75$$

Example 8.8

The CEs of the closed-loop control systems are provided as follows:

① $s^3 - 15s + 126 = 0$;

② $s^5 + 3s^4 - 3s^3 - 9s^2 - 4s - 12 = 0$

Find the numbers of roots on the RHS of the s-plane.

Solution (1)
The Routh array is

$$\begin{array}{lcc} s^3 & 1 & -15 \\ s^2 & 0 \approx \varepsilon & 126 \\ s^1 & \frac{-15\varepsilon-126}{\varepsilon} & 0 \\ s^0 & 126 & 0 \end{array}$$

$$\varepsilon \to 0 \Rightarrow \frac{-15\varepsilon - 126}{\varepsilon} \to -\infty$$

The signs of the first column changed twice. Hence, there are two roots on the RHS of the s-plane and the system is unstable.

Solution (2)
The Routh array is

$$\begin{array}{llll} s^5 & 1 & -3 & -4 \\ s^4 & 3 & -9 & -12 \\ s^3 & 0 & 0 & \\ s^2 & & & \\ s^1 & & & \\ s^0 & & & \end{array}$$

Implementing "auxiliary equation" by the s^4 row,

$$A(s) = 3s^4 - 9s^2 - 12 = 0$$

Taking derivative of $A(s)$ with respect to s, we get

$$\frac{dA(s)}{ds} = 12s^3 - 18s$$

The new Routh Array is

$$\begin{array}{llll} s^5 & 1 & -3 & -4 \\ s^4 & 3 & -9 & -12 \\ s^3 & 12 & -18 & \\ s^2 & -4.5 & -12 & \\ s^1 & -50 & & \\ s^0 & -12 & & \end{array}$$

The sign of the first column changes once. Therefore a root lies in the RHS of s-plane and the system is unstable. In addition, from solving the auxiliary equation, we can get two pairs of roots, a pair of the pure imaginary roots $s_{1,2} = \pm$ j and a pair of the pure real roots $s_{3,4} = \pm 2$.

8.4 Problems

P8.1. The open-loop TF of unit negative feedback system is as follows:

$$G(s)H(s)=\frac{100}{s^2(300s^2+600s+50)}$$

Determine whether the closed loop system is stable or not.

P8.2. Consider a unity feedback control system having the block diagram as shown in Fig. 8.8. Find the gain K that ensures the system to be stable.

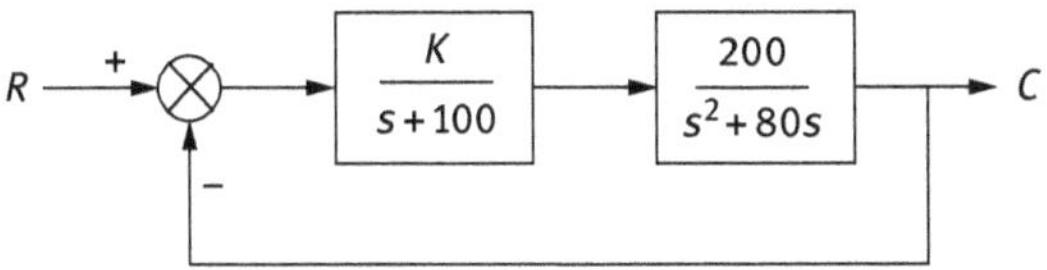

Fig. 8.8: The function block diagram for P 8.2.

P8.3. The CEs are as follow. How many roots are in the right part of the s-plane according to Routh-Hurwitz stability criterion?

$$(1)\ s^3-15s+126=0,\ \ (2)\ s^5+3s^4-3s^3-9s^2-4s-12=0$$

P8.4. The open-loop TF of the unit negative feedback system is

$$G(s)H(s)=\frac{k}{s(Ts+1)}$$

Please evaluate the value range of k and T if all characteristic roots locate on the left of the line $s=-a$.

9 Error Analysis and Calculation of Control Systems

In the last chapter, we introduced the stability of control systems. However, a control system does not always satisfy human demands, even if it is stable. Other pertinent factors, such as errors, must be taken into consideration. As the errors of the unsteady systems are inestimable, we only need to study the steady-state error for the stable feedback control systems, which is explained in detail in this chapter.

If disturbance is exerted on a balanced ball, it will eventually come back to the initial state. We call the ball system stable.

It belongs to the stability analysis of the control system, as shown in Fig. 9.1(a). In Fig. 9.1(b), the deviation will occur when the ball comes from one steady state to another or the ball comes back to balance after suffering some disturbance. We call the deviation as the steady-state error. The steady-state error is the core content in this chapter.

Control precision is a fundamental requirement, especially the steady-state error. It is an implication to interpret the accuracy of the control system. The steady-state error is closely related to the precision of all constituents, structure parameters of the system, and the input fed to the system.

9.1 Terminologies

9.1.1 Deviation

As Fig. 9.2 depicts, the action point of the deviation $E(s)$ is located at the summing point of the input signal and the primary feedback signal. The value of the deviation is the ratio of the input to the summation of the unit and the open-loop TF,

$$E(s) = X(s) - B(s) = X(s) - H(s) \cdot Y(s) = X(s) - H(s) \cdot \frac{G(s)}{1 + G(s)H(s)} \cdot X(s)$$

Simplifying the above equation, we have

$$E(s) = \frac{X(s)}{1 + G(s)H(s)} \tag{9.1}$$

9.1.2 Steady-State Deviation

According to the final value theorem on the expression of $E(s)$, the steady-state deviation $e_{ss}(t)$ of the system is

$$e_{ss}(t) = \lim_{t\to\infty} e(t) = \lim_{s\to 0} s \cdot E(s) = \lim_{s\to 0} s \cdot \frac{X(s)}{1 + G(s)H(s)} \tag{9.2}$$

https://doi.org/10.1515/9783110573275-009

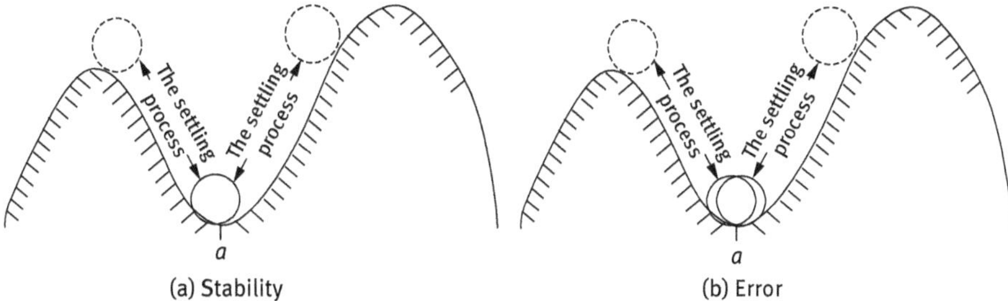

Fig. 9.1: The stability and the error of the control system.

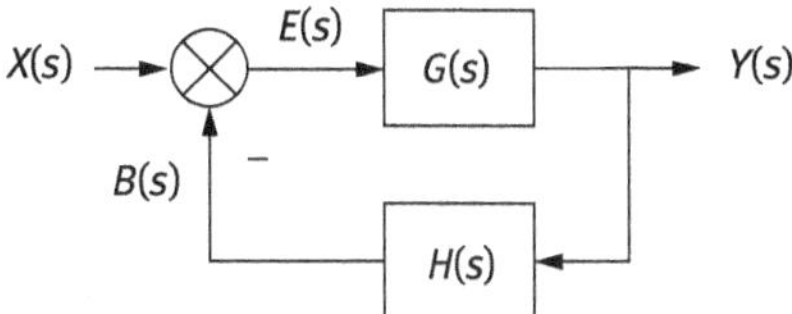

Fig. 9.2: The illustration for deviation.

9.1.3 Desired Output Value

When the deviation $E(s)$ is zero, the system will cease to adjust with the desired output value $Y_d(s)$. When the output variable $Y(s)$ is equal to the desired output variable $Y_d(s)$, the deviation $E(s)$ should be zero, such that

$$E(s) = X(s) - H(s)Y_d(s) = 0 \tag{9.3}$$

So, the desired output value $Y_d(s)$ will be

$$Y_d(s) = \frac{X(s)}{H(s)} \tag{9.4}$$

According to the above analysis, the theoretical definition of the desired value $Y_d(s)$ is the output value when no deviation occurs in the system. The physical definition of the desired value $Y_d(s)$ is the ideal situation of the control system. And the expression of the desired value $Y_d(s)$ is the input versus the TF of the feedback path.

9.1.4 Error

The difference between the desired output value $Y_d(s)$ and the actual output value $Y(s)$ is named as the error of the control system, where the signal of $\varepsilon(s)$ is

$$\varepsilon(s) = Y_d(s) - Y(s) = \frac{X(s)}{H(s)} - Y(s) \tag{9.5}$$

As

$$E(s) = X(s) - H(s) \cdot Y(s)$$

We have

$$\varepsilon(s) = \frac{E(s)}{H(s)} \tag{9.6}$$

Combining with eq. (9.1), we have

$$\varepsilon(s) = \frac{1}{1 + G(s)H(s)} \cdot \frac{X(s)}{H(s)} \tag{9.7}$$

The value of error is theoretically defined as the difference between the desired output value and the actual output value. You may note that when the system is a unit negative feedback system, eq. (9.5) changes to

$$\varepsilon(s) = X(s) - Y(s) \tag{9.8}$$

And eq. (9.7) changes to

$$\varepsilon(s) = \frac{X(s)}{1 + G(s)} \tag{9.9}$$

At the same time, eq. (9.1) could also be changed to

$$E(s) = X(s) - Y(s) = \frac{X(s)}{1 + G(s)} \tag{9.10}$$

Obviously, when the system is a unit feedback system, the error is equal to the deviation. In other words, when the system is a unit feedback system, the error is the input minus output, which is also equal to the deviation of the system shown by the doted lines in Fig. 9.3.

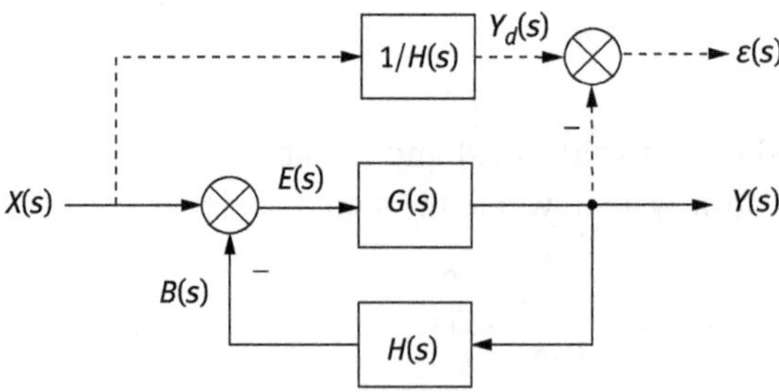

Fig. 9.3: The relationships among $\varepsilon(s)$, $E(s)$, $Y(s)$, and $Y_d(s)$.

9.1.5 Steady-State Error

According to the final value theorem, we can derive the steady-state error $\varepsilon_{ss}(t)$ from the error $\varepsilon(s)$,

$$\varepsilon_{ss}(t) = \lim_{t\to\infty} \varepsilon(t) = \lim_{s\to 0} s \cdot \varepsilon(s) = \lim_{s\to 0} s \cdot \frac{1}{1+G(s)H(s)} \cdot \frac{X(s)}{H(s)} \tag{9.11}$$

According to the relationship between the error and the deviation, the steady-state error in eq. (9.6) can also be gained by the final value theorem,

$$\varepsilon_{ss}(t) = \lim_{s\to 0} s \cdot \varepsilon(s) = \lim_{s\to 0} s \cdot \frac{E(s)}{H(s)} = e_{ss} \cdot \lim_{s\to 0} \frac{1}{H(s)} = \frac{e_{ss}}{H(0)} \tag{9.12}$$

Example 9.1

A speed regulating system is shown in Fig. 9.4. Find its steady-state error $\varepsilon_{ss}(t)$ when the input voltage is 10 V.

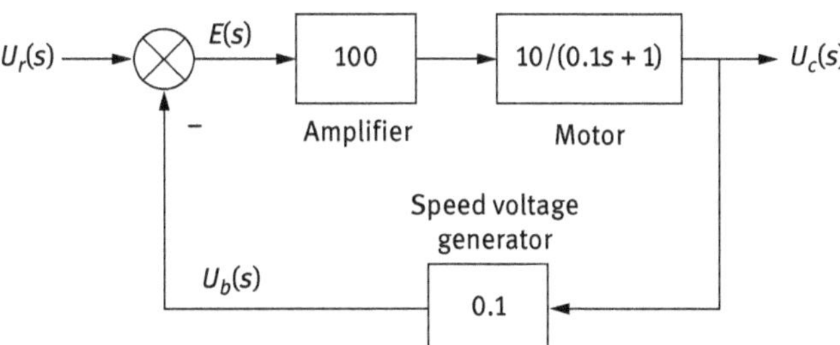

Fig. 9.4: Function block diagram for example 9.1.

Solution

From the function block diagram, the forward path TF is obtained as

$$G(s) = 100 \times \frac{10}{0.1s+1}$$

The feedback path TF is

$$H(s) = 0.1$$

The input signal is

$$X(s) = \frac{10}{s}$$

Hence, substituting these equations into eq. (9.11), we could work out the steady-state error as

$$\varepsilon_{ss}(t) = \lim_{s\to 0} s \cdot \frac{1}{1+100\times\frac{10}{0.1s+1}\times 0.1}\times\frac{\frac{10}{s}}{0.1} = \frac{100}{101} = 0.99$$

9.2 Static Error Coefficients

9.2.1 Two Impact Factors of the Steady-State Error

Using the given condition for unit negative feedback, eq. (9.11) can be categorized by two impact factors, namely the open-loop TF $G(s)$, which represents the system type, and the input variable $X(s)$.

9.2.1.1 System Type

As a matter of fact, the open-loop TF $G(s)$ represents the system type. The open-loop TF of the unit negative feedback system is

$$G(s) = \frac{K(\tau_1 s+1)(\tau_2 s+1)\cdots(\tau_m s+1)}{s^N(T_1 s+1)(T_2 s+1)\cdots(T_n s+1)} \tag{9.13}$$

According to eq. (9.13), the system can be classified as 0, I, II, etc. depending on the numbers of the corresponding integral link. For example, if $N = 0$, the type of this system is 0, while if $N = 1$, the type of this system is 1, and so on. When the system stability is concerned, the "type II +" system is seldom considered in the practical control systems.

9.2.1.2 Input Signal Type

The input variable $X(s)$ represents the input signal type. The most common input signals are step, ramp, and acceleration signals given by

$$x_s(t) = R,\ x_r(t) = Rt,\ x_a(t) = \frac{Rt^2}{2}$$

Their Laplace transforms are

$$X_s(s) = \frac{R}{s},\ X_r(s) = \frac{R}{s^2},\ X_a(s) = \frac{R}{s^3}$$

Hence, we can derive the steady-state error of the system according to the open-loop TF $G(s)$ and the input variable $X(s)$.

9.2.2 The Static Error Coefficients and the Steady-State Error

There are three types of steady-state errors corresponding to three different input signals for a unit negative feedback system.

9.2.2.1 Step Input Signal

According to eq. (9.11),

$$\varepsilon_{ss1} = \lim_{s\to 0} s \cdot E(s) = \lim_{s\to 0} s \cdot \frac{X(s)}{1+G(s)} = \lim_{s\to 0} s \cdot \frac{1}{1+G(s)} \cdot \frac{R}{s} = \frac{R}{1+\lim\limits_{s\to 0} G(s)} \tag{9.14}$$

A new terminology, static position error coefficient K_p, is defined as

$$K_p = \lim_{s\to 0} G(s) \tag{9.15}$$

Hence, the steady-state error with the step input signal is

$$\varepsilon_{ss1} = \frac{R}{1+K_p} \tag{9.16}$$

Now we should discuss three situations within various system types.

(1) Type 0 ($N = 0$).
The static position error coefficient is

$$K_p = \lim_{s\to 0} G(s) = \lim_{s\to 0} \frac{K(\tau_1 s+1)(\tau_2 s+1)\cdots(\tau_m s+1)}{(T_1 s+1)(T_2 s+1)\cdots(T_n s+1)} = K$$

The steady-state error with the step input signal and system type 0 is

$$\varepsilon_{ss1} = \frac{R}{1+K} \tag{9.17}$$

(2) Type I ($N = 1$).
The static position error coefficient is

$$K_p = \lim_{s\to 0} G(s) = \lim_{s\to 0} \frac{K(\tau_1 s+1)(\tau_2 s+1)\cdots(\tau_m s+1)}{s(T_1 s+1)(T_2 s+1)\cdots(T_n s+1)} = \infty$$

The steady-state error with the step input signal and system type I is

$$\varepsilon_{ss1} = 0 \tag{9.18}$$

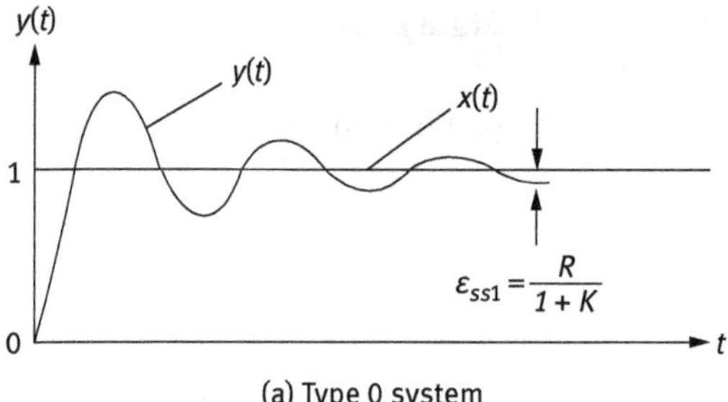

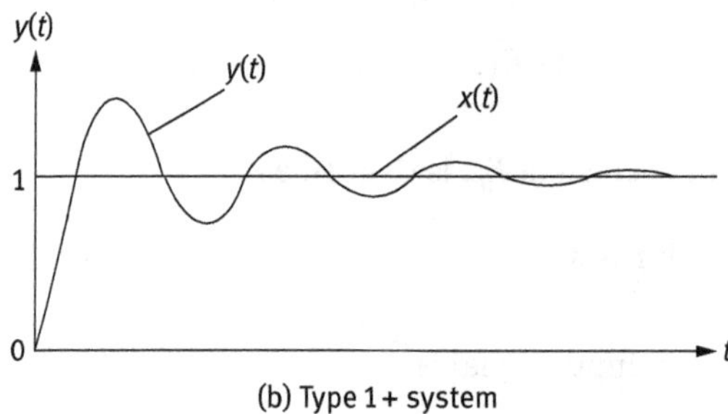

Fig. 9.5: Step response curves of the unit negative feedback control system.

(3) Type II (N=2).
The static position error coefficient is

$$K_p = \lim_{s\to 0} G(s) = \lim_{s\to 0} \frac{K(\tau_1 s+1)(\tau_2 s+1)\cdots(\tau_m s+1)}{s^2(T_1 s+1)(T_2 s+1)\cdots(T_n s+1)} = \infty$$

The steady-state error with the step input signal and system type II is

$$\varepsilon_{ss1} = 0 \tag{9.19}$$

The step response curve of the unit negative feedback system is shown in Fig. 9.5. The left hand side of the figure is for "type 0," while the right hand side is for "type I +." In terms of the expression $\varepsilon = R/(1+K)$ for "type 0," the steady-state error ε should be sufficiently small if the open-loop gain coefficient K is big enough (overlarge coefficient will make the system unstable). According to Fig. 9.5(b), the system type should be "I +" if you want to get zero steady-state error with the step input signal.

9.2.2.2 Ramp Input Signal

According to eq. (9.11), together with the ramp input signal,

$$\varepsilon_{ss2} = \lim_{s\to 0} s\cdot E(s) = \lim_{s\to 0} s\cdot \frac{X(s)}{1+G(s)} = \lim_{s\to 0} s\cdot \frac{1}{1+G(s)}\cdot\frac{R}{s^2} = \frac{R}{\lim_{s\to 0} sG(s)} \tag{9.20}$$

Now, a neo-terminology, static velocity error coefficient K_v, is presented here as

$$K_v = \lim_{s\to 0} sG(s) \tag{9.21}$$

Hence, the steady-state error with the ramp input signal is

$$\varepsilon_{ss2} = \frac{R}{K_v} \tag{9.22}$$

Let us discuss three situations with respect to various system types.

(1) Type 0 ($N = 0$).
The static velocity error coefficient is

$$K_v = \lim_{s\to 0} sG(s) = \lim_{s\to 0} s \frac{K(\tau_1 s+1)(\tau_2 s+1)\cdots(\tau_m s+1)}{(T_1 s+1)(T_2 s+1)\cdots(T_n s+1)} = 0$$

The steady-state error with the ramp input signal and system type 0 is

$$\varepsilon_{ss2} = \infty \tag{9.23}$$

(2) Type I ($N = 1$).
The static velocity error coefficient is

$$K_v = \lim_{s\to 0} sG(s) = \lim_{s\to 0} s \frac{K(\tau_1 s+1)(\tau_2 s+1)\cdots(\tau_m s+1)}{s(T_1 s+1)(T_2 s+1)\cdots(T_n s+1)} = K$$

The steady-state error with the ramp input signal and system type I is

$$\varepsilon_{ss2} = \frac{R}{K} \tag{9.24}$$

(3) Type II ($N = 2$).
The static velocity coefficient is

$$K_v = \lim_{s\to 0} sG(s) = \lim_{s\to 0} s \frac{K(\tau_1 s+1)(\tau_2 s+1)\cdots(\tau_m s+1)}{s^2(T_1 s+1)(T_2 s+1)\cdots(T_n s+1)} = \infty$$

The steady-state error with the ramp input signal and system type II is

$$\varepsilon_{ss2} = 0 \tag{9.25}$$

According to Fig. 9.6(a), the system type 0 cannot follow the ramp input. As per Fig. 9.6(b), the system type I can follow the ramp input with the steady-state error. And in terms of the expression of $\varepsilon = R/K$, the larger the open-loop gain K, the smaller the steady-state error. For Fig. 9.6(c), "type II +" can follow the ramp input signal precisely without any steady-state error.

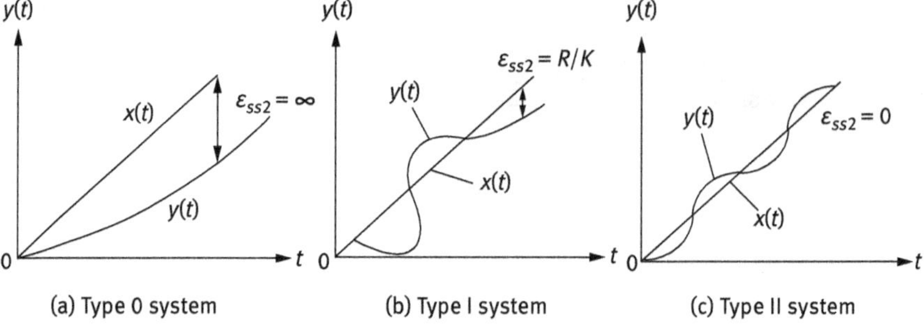

Fig. 9.6: Ramp response curves of unit negative feedback control system.

9.2.2.3 Acceleration Input Signal

According to eq. (9.11),

$$\varepsilon_{ss3} = \lim_{s\to 0} s \cdot E(s) = \lim_{s\to 0} s \cdot \frac{X(s)}{1+G(s)} = \lim_{s\to 0} s \cdot \frac{1}{1+G(s)} \cdot \frac{R}{s^3} = \frac{R}{\lim_{s\to 0} s^2 G(s)} \tag{9.26}$$

A new terminology, static acceleration error coefficient K_a, is introduced here as

$$K_a = \lim_{s\to 0} s^2 G(s) \tag{9.27}$$

Hence, the steady-state error generated by acceleration input signal is

$$\varepsilon_{ss3} = \frac{R}{K_a} \tag{9.28}$$

Now we should discuss three situations according to distinct system types.

(1) Type 0 ($N = 0$).
The static acceleration error coefficient is

$$K_a = \lim_{s\to 0} s^2 G(s) = \lim_{s\to 0} s^2 \frac{K(\tau_1 s+1)(\tau_2 s+1)\cdots(\tau_m s+1)}{(T_1 s+1)(T_2 s+1)\cdots(T_n s+1)} = 0$$

The steady-state error caused by acceleration input signal and system type 0 is

$$\varepsilon_{ss3} = \infty \tag{9.29}$$

(2) Type I ($N = 1$).
The static acceleration error coefficient is

$$K_a = \lim_{s\to 0} s^2 G(s) = \lim_{s\to 0} s^2 \frac{K(\tau_1 s+1)(\tau_2 s+1)\cdots(\tau_m s+1)}{s(T_1 s+1)(T_2 s+1)\cdots(T_n s+1)} = 0$$

The steady-state error by acceleration input signal and system type I is

$$\varepsilon_{ss3} = \infty \tag{9.30}$$

(3) Type II ($N = 2$).

The static acceleration coefficient is

$$K_a = \lim_{s\to 0} s^2 G(s) = \lim_{s\to 0} s^2 \frac{K(\tau_1 s+1)(\tau_2 s+1)\cdots(\tau_m s+1)}{s^2(T_1 s+1)(T_2 s+1)\cdots(T_n s+1)} = K$$

The steady-state error with the acceleration input signal and system type II is

$$\varepsilon_{ss3} = \frac{R}{K} \tag{9.31}$$

According to eqs. (9.29 and 9.30), the system type 0 and I do not follow the acceleration input signal. From eq. (9.31), we can say that the system type II and beyond can be followed by the steady-state error, shown in Fig. 9.7. In terms of the expression of $\varepsilon = R/K$, however, its value is inversely proportional to the open-loop gain coefficient K.

Table 9.1 demonstrates the steady-state errors of all three typical inputs with various open-loop types. The equations for the steady-state errors can be applied directly without deduction. In those equations for the steady-state errors, R

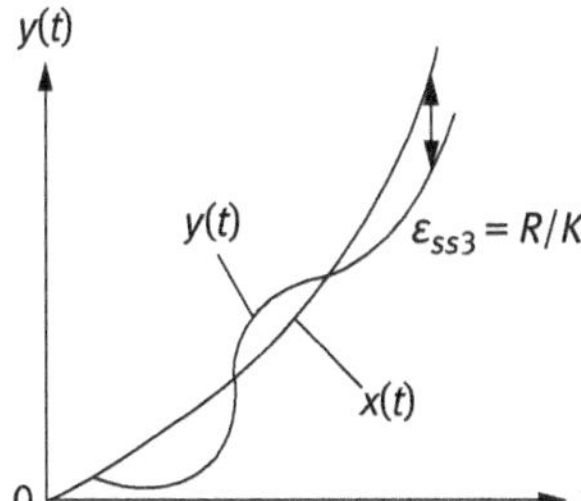

Fig. 9.7: Acceleration response curve of unit feedback negative control system.

Table 9.1: Steady-state errors of three typical input signals by various system types

System type	Typical input signal		
	Step	**Ramp**	**Acceleration**
	$x(t) = R$ $X(s) = \frac{R}{s}$	$x(t) = Rt$ $X(s) = \frac{R}{s^2}$	$x(t) = \frac{R}{2}t^2$ $X(s) = \frac{R}{s^3}$
0	$\frac{R}{1+K}$	∞	∞
I	0	$\frac{R}{K}$	∞
II	0	0	$\frac{R}{K}$

represents the amplitude of the input (note the situation of acceleration input) and K is the open-loop gain of the time constant type.

Example 9.2
A second-order oscillation system is shown in Fig. 9.8. Evaluate the steady-state errors when the input signals are unit step signal, unit ramp signal, and unit acceleration signal.

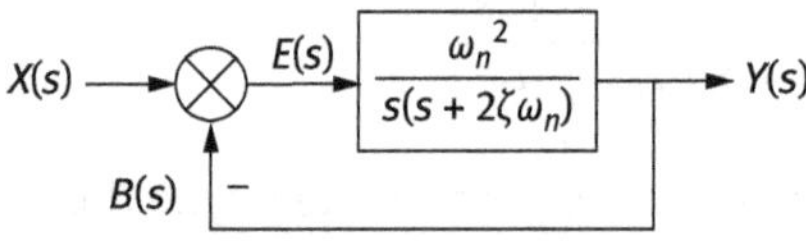

Fig. 9.8: Function block diagram for example 9.2.

Solution
According to the function block diagram, the open-loop TF with the "+1" form is

$$G(s)H(s) = \frac{\frac{\omega_n}{2\zeta}}{s\left(\frac{s}{2\zeta\omega_n}+1\right)}$$

Hence, the system type is I and its open-loop gain coefficient is

$$K = \frac{\omega_n}{2\zeta}$$

According to Table 9.1, when the input is unit step signal, the steady-state error is

$$\varepsilon_{ss1} = 0$$

The steady-state error when the input signal is the unit ramp signal is

$$\varepsilon_{ss2} = \frac{R}{K} = \frac{1}{K} = \frac{2\zeta}{\omega_n}$$

When the input signal is the unit acceleration signal, its steady-state error is

$$\varepsilon_{ss3} = \infty$$

9.3 Steady-State Error Calculation

In order to analyze the function block diagram, we could first perform the steady-state deviation calculation. And then, referring to the relationship between the deviation and error, we could get the exactly value for the steady-state error.

9.3.1 Steady-State Deviation Calculation

If the system is subject to the input signal $X(s)$ and the disturbance signal $N(s)$ simultaneously, as shown in Fig. 9.9, the deviations introduced by the two signals should be evaluated by superposition of the two input signals,

$$E(s) = E_X(s) + E_N(s) \tag{9.32}$$

According to the definition of the deviation $E(s)$, the action point of the total deviation $E(s)$ is the summing point of the input variable and primary feedback variable.

9.3.1.1 Determination of $E_X(s)$

In order to evaluate the deviation introduced by the input signal $X(s)$, we should first assume $N(s) = 0$. Thus, Fig. 9.9 is changed to Fig. 9.10. The $E_X(s)$ subject to the single input signal $X(s)$ is

$$\begin{aligned} E_X(s) &= X(s) - H(s) \cdot Y_X(s) \\ &= X(s) - H(s) \cdot \frac{G_1(s)G_2(s)}{1 + G_1(s)G_2(s)H(s)} \cdot X(s) \\ &= \frac{X(s)}{1 + G_1(s)G_2(s)H(s)} \end{aligned} \tag{9.33}$$

According to the final value theorem, at this time, the steady-state deviation is

$$e_{ssx} = \lim_{s \to 0} sE_X(s) = \lim_{s \to 0} s \cdot \frac{X(s)}{1 + G_1(s)G_2(s)H(s)} \tag{9.34}$$

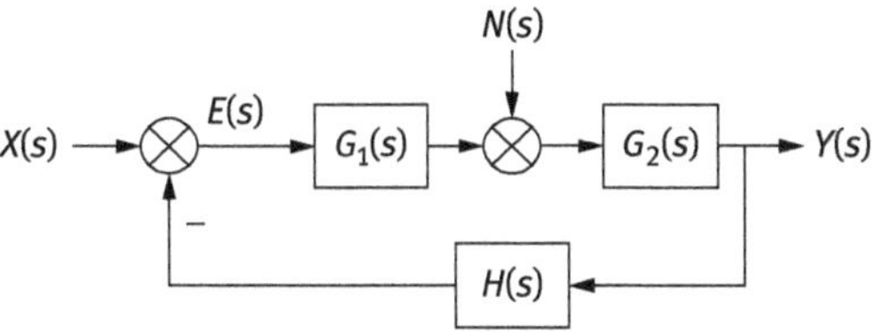

Fig. 9.9: Function block diagram with input signal $X(s)$ and disturbance signal $N(s)$.

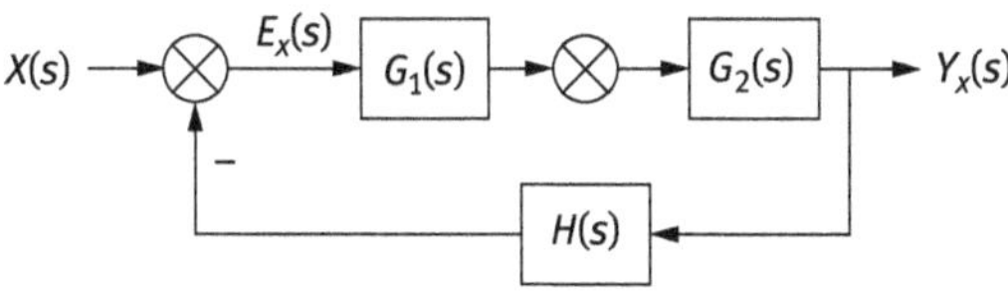

Fig. 9.10: Function block diagram with input signal $X(s)$.

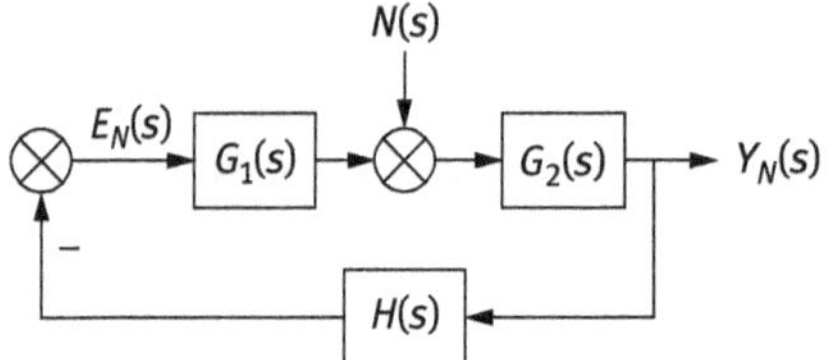

Fig. 9.11: Function block diagram with disturbance signal $N(s)$.

9.3.1.2 Determination of $E_N(s)$

In order to evaluate the deviation caused by disturbance signal $N(s)$, it is assumed that $X(s) = 0$. Hence, Fig. 9.9 is altered by Fig. 9.11. The $E_N(s)$ subject to the disturbance input $N(s)$ is

$$E_N(s) = -Y_N(s)H(s) = -\frac{G_2(s)H(s)}{1+G_1(s)G_2(s)H(s)} \cdot N(s) \tag{9.35}$$

By the final value theorem, we can get the steady-state deviation,

$$e_{ssn} = \lim_{s\to 0} sE_N(s) = \lim_{s\to 0} s\left[\frac{-G_2(s)H(s)}{1+G_1(s)G_2(s)H(s)}N(s)\right] \tag{9.36}$$

9.3.1.3 Determination of $E(s)$

The total deviation by the input signal $X(s)$ and the disturbance signal $N(s)$ is

$$E(s) = E_X(s) + E_N(s) = \frac{X(s) - G_2(s)H(s)N(s)}{1+G_1(s)G_2(s)H(s)} \tag{9.37}$$

The total steady-state deviation is

$$\begin{aligned} e_{ss} = e_{ssx} + e_{ssn} &= \lim_{s\to 0} sE_X(s) + \lim_{s\to 0} sE_N(s) \\ &= \lim_{s\to 0} s \cdot \frac{X(s)}{1+G_1(s)G_2(s)H(s)} + \lim_{s\to 0} s\left[\frac{-G_2(s)H(s)}{1+G_1(s)G_2(s)H(s)}N(s)\right] \end{aligned} \tag{9.38}$$

9.3.2 Steady-State Error Calculation

According to the relationship between the steady-state deviation and the steady-state error established in eq. (9.12), we can get the steady-state error generated by the input signal $X(s)$ as

$$\varepsilon_{ssx} = \lim_{s\to 0} s\frac{E_X(s)}{H(s)} = \lim_{s\to 0} s \cdot \frac{X(s)}{1+G_1(s)G_2(s)H(s)} \cdot \frac{1}{H(s)} \tag{9.39}$$

At the same time, steady-state error introduced by the input signal $N(s)$ is

$$\begin{aligned}\varepsilon_{ssn} &= \lim_{s\to 0} s\frac{E_N(s)}{H(s)} = \lim_{s\to 0} s\left[\frac{-G_2(s)H(s)}{1+G_1(s)G_2(s)H(s)}N(s)\right]\cdot\frac{1}{H(s)} \\ &= \lim_{s\to 0} s\cdot\frac{-G_2(s)N(s)}{1+G_1(s)G_2(s)H(s)}\end{aligned} \tag{9.40}$$

Hence, the total steady-state error by $X(s)$ and $N(s)$ is

$$\varepsilon_{ss} = \varepsilon_{ssx} + \varepsilon_{ssn} = \lim_{s\to 0} s\cdot\frac{X(s)}{1+G_1(s)G_2(s)H(s)}\cdot\frac{1}{H(s)} + \lim_{s\to 0} s\cdot\frac{-G_2(s)N(s)}{1+G_1(s)G_2(s)H(s)} \tag{9.41}$$

Example 9.3

An electro-hydraulic servo valve control system is shown in Fig. 9.12 and the input signal is $i(t)$ = 0.01A. The Baffle angular displacement caused by thermal deformation is a disturbance signal, which is $f(t)$ = −0.000314 rad. Find out the steady-state error caused by the two input signals.

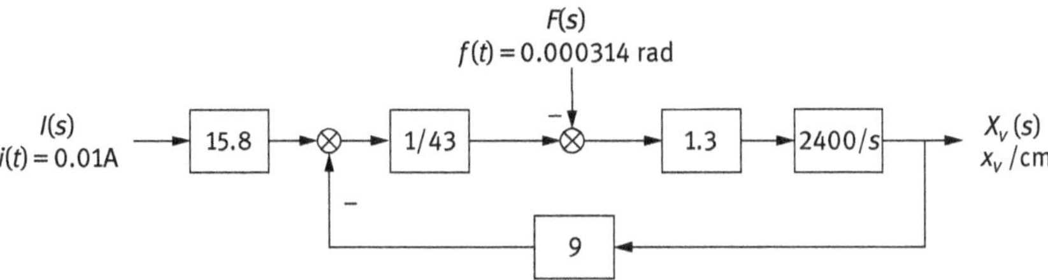

Fig. 9.12: Function block diagram for example 9.3.

Solution

First, we should make some simplification of the input signal,

$$I'(s) = \frac{0.01\times 15.8}{s} = \frac{0.158}{s}$$

So, according to equations 9.33–9.41 and Fig. 9.12, we have the total steady-state error,

$$\begin{aligned}\varepsilon_{ss}(t) &= \varepsilon_{ssI'}(t) + \varepsilon_{ssF}(t) \\ &= \lim_{s\to 0} s\cdot[I'(s) - H(s)\cdot X_{VI'}(s)]\cdot\frac{1}{H(s)} + \lim_{s\to 0} s\cdot[-H(s)\cdot X_{VF}(s)]\cdot\frac{1}{H(s)}\end{aligned} \tag{9.42}$$

where $H(s)$ = 1. According to Fig. 9.12, we can get the output signal $X_{VI'}(s)$ introduced by the input signal $I'(s)$ as

$$X_{VI'}(s) = \frac{\frac{1}{43}\times 1.3\times\frac{2400}{s}}{1+\frac{1}{43}\times 1.3\times\frac{2400}{s}\times 9}\times\frac{0.158}{s}$$

And the output signal $X_{VF}(s)$ introduced by the disturbance input signal $F(s)$ is

$$X_{VF}(s) = \frac{1.3 \times \frac{2400}{s}}{1 + \frac{1}{43} \times 1.3 \times \frac{2400}{s} \times 9} \times \left(-\frac{0.000314}{s}\right)$$

Substituting the above equations into eq. (9.42),

$$\begin{aligned}\varepsilon_{ss}(t) &= \varepsilon_{ssI'}(t) + \varepsilon_{ssF}(t)\\ &= \lim_{s\to 0} s\cdot[I'(s) - H(s)\cdot X_{VI'}(s)]\cdot\frac{1}{H(s)} + \lim_{s\to 0} s\cdot[-H(s)\cdot X_{VF}(s)]\cdot\frac{1}{H(s)}\\ &= 0.015\,\text{mm}\end{aligned}$$

9.4 Methods for Reducing Steady-State Error

In most practical engineering systems, the steady-state error is unavoidable. But we could always find a proper way to reduce the steady-state error. In this section, we will introduce three useful methods to reduce the steady-state error by the disturbance signal. Besides considering the error due to the disturbance signal, we will also explain a method to reduce the steady-state error caused by the feedback loop control system itself.

9.4.1 Increase Open-Loop Gain

According to Table 9.1, we conclude that the steady-state error will decrease when open-loop gain increases for cases such that type 0 system follows the step input, type I system follows the ramp input, and type II system follows the acceleration input. However, overlarge open-loop gain will weaken the stability of the system (see Example 8.4).

Example 9.4

Find the steady-state error ε_{ssN} introduced by the disturbance input signal $N(s)$ for the system shown in Fig. 9.13. The given conditions are

$$N(s) = \frac{1}{s}, \quad G_1(s) = K_1, \quad G_2(s) = \frac{K_2}{s(T_2 s + 1)}$$

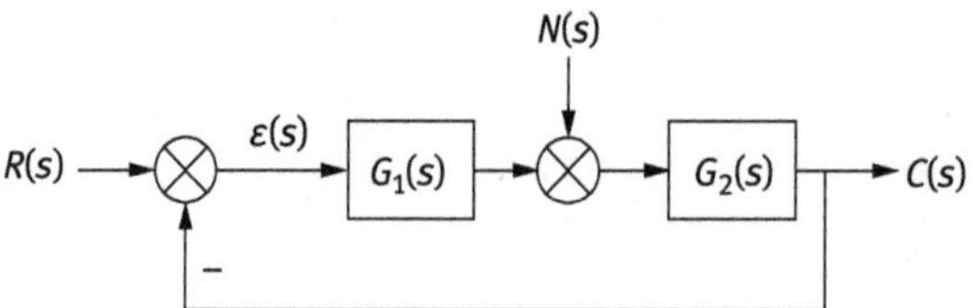

Fig. 9.13: Function block diagram for example 9.4.

Solution

The steady-state error ε_{ssN} introduced by the disturbance signal $N(s)$ is

$$\varepsilon_{ssN} = \lim_{s\to 0} s\frac{-G_2(s)}{1+G_1(s)G_2(s)}N(s)$$

$$= \lim_{s\to 0}\left[s\cdot\frac{-\frac{K_2}{s(T_2s+1)}}{1+K_1\frac{K_2}{s(T_2s+1)}}\cdot\frac{1}{s}\right]$$

$$= \lim_{s\to 0}\left[-\frac{K_2}{s(T_2s+1)+K_1K_2}\right] = -\frac{1}{K_1}$$

According to example 9.4, we can draw a general conclusion: the open-loop gain K_1 of controller $G_1(s)$ is inversely proportional to the steady-state error ε_{ssN} introduced by the disturbance signal $N(s)$.

9.4.2 Increasing System Types

According to Table 9.1, you will find that for the same input signal, the steady-state error will become smaller when the types of the system become complex. However, more integral links will turn the system less stable.

Example 9.5

Find the steady-state error ε_{ssN} introduced by the disturbance signal $N(s)$ to the system shown in Fig. 9.14. The given conditions are

$$N(s) = \frac{1}{s},\quad G_1(s) = K_1\left(1+\frac{1}{T_1s}\right),\quad G_2(s) = \frac{K_2}{s(T_2s+1)}$$

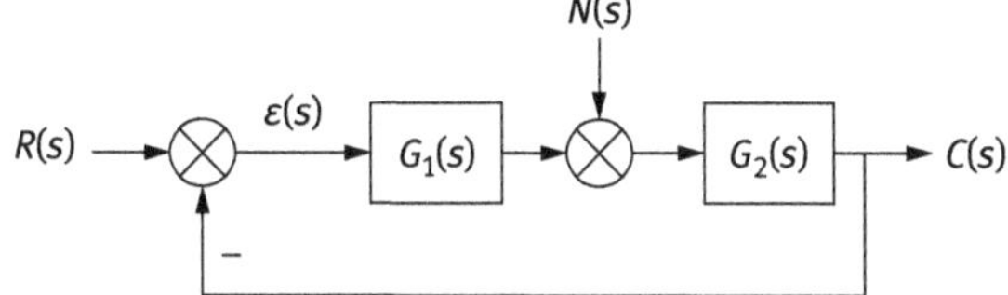

Fig. 9.14: Function block diagram for example 9.5.

Solution

The steady-state error ε_{ssN} introduced by the disturbance signal $N(s)$ is

$$\varepsilon_{ssN} = \lim_{s\to 0} s\frac{-G_2(s)}{1+G_1(s)G_2(s)}N(s)$$

$$= \lim_{s\to 0}\left[s\cdot\frac{-\frac{K_2}{s(T_2s+1)}}{1+K_1(1+\frac{1}{T_1s})\cdot\frac{K_2}{s(T_2s+1)}}\right]\cdot\frac{1}{s}$$

$$= \lim_{s\to 0}\left[-\frac{K_2T_1s}{T_1s^2(T_2s+1)+K_1K_2(T_1s+1)}\right] = 0$$

Comparing with the above two examples, we can arrive at some meaningful conclusions.

(1) The number of the integral link in the forward path in Example 9.5 is more than that in Example 9.4.

(2) The steady-state error in Example 9.5 is zero, but that in Example 9.4 is a nonzero constant.

(3) Proper increase in the type of the system can reduce its steady-state error.

9.4.3 Feed Forward Control

In addition to the previous methods to eliminate the steady-state error caused by the disturbance signal *t*, one can employ the feed forward control. For the system shown in Fig. 9.15(a), with the disturbance signal $N(s)$, we can adopt a compensation link $G_F(s)$ from the disturbance signal $N(s)$ to the input signal $X(s)$, shown in Fig. 9.15(b). This method is called the feed forward control method. According to Fig. 9.15 (b),

$$\begin{cases} Y(s) = G_2(s)[N(s) + E(s)G_1(s)] & (9.43a) \\ E(s) = X(s) - Y(s) - G_F(s)N(s) & (9.43b) \end{cases}$$

Substituting eq. (9.43b) into eq. (9.43a),

$$Y(s) = Y_X(s) + Y_N(s) = \frac{G_1(s)G_2(s)}{1 + G_1(s)G_2(s)}X(s) + \frac{[1 - G_1(s)G_F(s)]G_2(s)}{1 + G_1(s)G_2(s)}N(s)$$

From the above equation, we know that the output of the system is determined by the input $X(s)$ and the disturbance $N(s)$. In order to obliterate the impact of disturbance $N(s)$, let

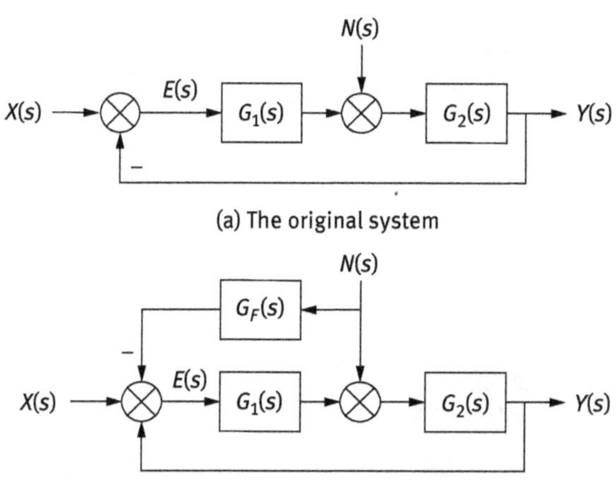

Fig. 9.15: Feed forward control method.

$$Y_N(s) = 0$$

That is,

$$\frac{[1 - G_1(s)G_F(s)]G_2(s)}{1 + G_1(s)G_2(s)} N(s) = 0$$

Obviously,

$$1 - G_1(s)G_F(s) = 0$$

Hence,

$$G_F(s) = \frac{1}{G_1(s)}$$

Thus, if you want to obliterate the impact of the disturbance signal, you could introduce a compensation link from the disturbance signal to the input signal with the expression of reciprocal of $G_1(s)$, where $G_1(s)$ is the forward path TF from the input signal to the summing point of the disturbance signal adding in.

9.4.4 Compound Control

How to reduce the steady-state error introduced by the closed-loop control system itself (not by some disturbance input)? For the closed-loop control system, shown in Fig. 9.16(a), introducing a compensation link $G_c(s).G_c(s)$ generates the compound control with the deviation signal $E(s)$ in order to reduce the steady-state error, shown in Fig. 9.16(b).

The closed-loop TF of Fig. 9.16(b) is

$$\frac{Y(s)}{X(s)} = \frac{[G_1(s) + G_c(s)]G_2(s)}{1 + G_1(s)G_2(s)}$$

Rearranging the above equation,

$$Y(s) = \frac{G_1(s)G_2(s) + G_c(s)G_2(s)}{1 + G_1(s)G_2(s)} X(s)$$

For this unit negative feedback system, the error is

$$E(s) = X(s) - Y(s) = X(s) - \frac{G_1(s)G_2(s) + G_c(s)G_2(s)}{1 + G_1(s)G_2(s)} X(s) = \frac{1 - G_c(s)G_2(s)}{1 + G_1(s)G_2(s)} X(s)$$

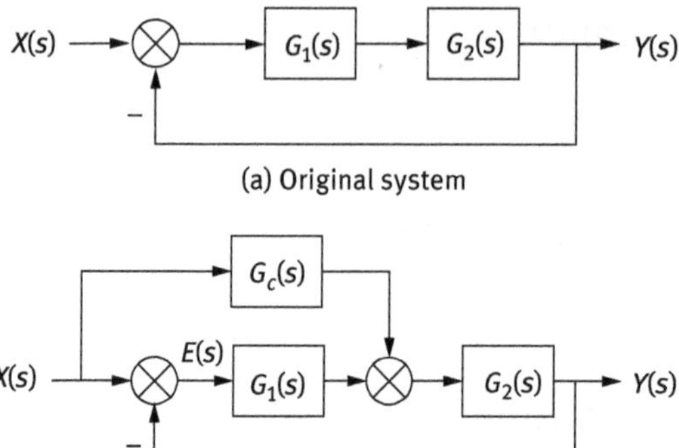

(a) Original system

(b) Compensation link with the compound control $G_c(s)$

Fig. 9.16: Compound control method.

In order to satisfy $E(s) = 0$, let

$$1 - G_c(s)G_2(s) = 0$$

Hence,

$$G_c(s) = \frac{1}{G_2(s)}$$

Based on the above analysis, the error will be zero if a proper compensation link $G_c(s)$ is selected, i.e., $Y(s) = X(s)$. It also means that the output signal reproduces the input signal or the output signal is the desired input signal. Therefore, compensation control is also named the output invariant condition considering the given input.

9.5 Problems

P9.1. The open-loop TF for the unit negative feedback system is

$$G(s)H(s) = \frac{10}{s(s+1)}$$

Find:

(1) The static error coefficients K_p, K_v, K_a.

(2) The steady-state error ε_{ss} with the input signal $r(t) = 2 + t + 2t^2$.

P9.2. The two control systems shown in Fig. 9.17 have the unit step input signal. Evaluate:

(1) For Fig. 9.17(a), the steady-state error $\varepsilon_{ss}(t)$ when the open-loop gain $K_c = 1/K$.

(2) For Fig. 9.17(b), the steady-state error $\varepsilon_{ss}(t)$ when the closed-loop gain $K_pK \gg 1$.

(3) Comparing the precision of those two control systems under the conditions that the gain K increases by 10%, together with $K_c = 1/K$ and $K_pK = 100$ because of the environmental change and the aging of the components.

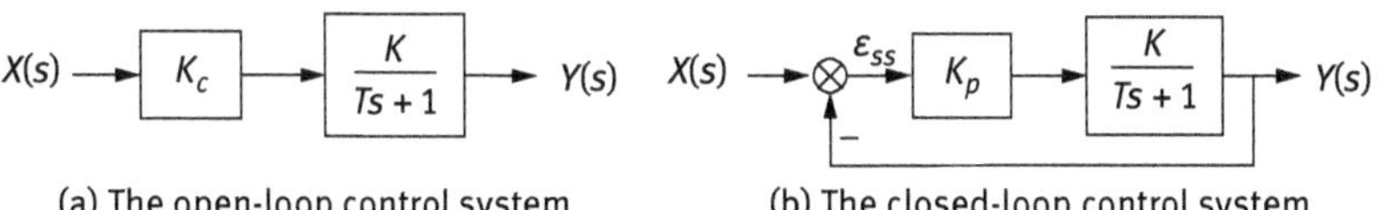

(a) The open-loop control system (b) The closed-loop control system

Fig. 9.17: Founction diagrams for P9.2

P9.3. A compound control system is shown in Fig. 9.18, in which $K_1 = 2$, $K_2 = 2$, $\zeta = 0.5$, the input signal is the unit ramp signal,

(1) Find the steady-state error ε_{ss} when $G_3(s) = 0$.

(2) Find the value of τ when compound control $G_3(s) = \tau s$ in order to get $\varepsilon_{ss} = 0$.

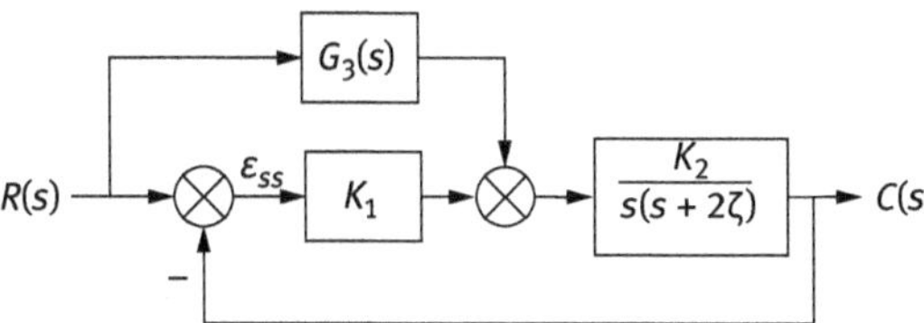

Fig. 9.18: The compound control system.

References

[1] Li, L. J.; Hu, Y. J.; Xiao, Y. M. Control Fundamentals of Mechanical Engineering [M]. Beijing: China Machine Press. 2013.

[2] Zhu, J. B.; Xu, X. L.; Chen, X. M. et al. Control Fundamentals of Mechanical Engineering. 2nd edn [M]. Beijing: China Machine Press. 2013.

[3] Dong, J. X.; Zhao, C. D.; Xiong, S. S. et al. Introduction to Control Engineering. 2nd edn [M]. Beijing: Tsinghua University Press. 2006.

[4] Vincent Del Toro; Sydney R. Parker. Principles of Control Systems Engineering [M]. New York: McGraw-Hill Book Company, INC. 1960.

[5] Zhang, S. C. Introduction to Control Engineering [M]. Hangzhou: Zhejiang University Press. 1991.

[6] Xu, X. L.; Wang, C. L. Introduction to Control Engineering [M]. Beijing: National Defense Industry Press. 2008.

[7] Cho, W. S. To. Introduction to and Dynamics and Control in Mechanical Engineering Systems [M]. West Sussex: John Willey & Sons, Ltd,. 2016.

[8] Hu, S. S. Principles of Automatic Control. 4th edn [M]. Beijing: Science Press. 2001.

[9] Charles, M. Close; Dean, H. Frederich; Jonathan, C.Newell. Modeling and Analysis of Dynamic Systems. 3rd edn [M]. New York: John Willey & Sons, Inc. 2001.

[10] Huang, A. Y.; Li, W. G.; Yang, Z. Z. Control Fundamentals of Mechanical Engineering [M]. Wuhan: Wuhan University of Technology Press. 2008.

[11] Zhu, S. X.; Xing, Y. J.; Han, L. Y. Control Fundamentals of Mechanical Engineering [M]. Beijing: Tsinghua University Press. 2008.

[12] Kong, X. D.; Wang, Y. Q.; Gao, Y. J. Introduction to Control Engineering [M]. Beijing: Machinery Industry Press. 2008.

[13] Dong, Y. H.; Xu, L. P. Control Fundamentals of Mechanical Engineering. 2nd edn [M]. Beijing: Machinery Industry Press. 2015.

[14] Wang, J. W.; Wu, Z.S. Introduction to Control Engineering [M]. Beijing: Higher Education Press. 2001.

[15] Xiong, L. C.; Yang, K. C.; Wu, B. Control Fundamentals of Mechanical Engineering Learning Guidance and Solutions. 4th edn [M]. Wuhan: Huazhong University of Science and Technology Press. 2000.

[16] Xiong, L. C.; Yang, K. C.; Wu, B. Control Fundamentals of Mechanical Engineering Learning Guidance and Solutions. (revised edition) [M]. Wuhan: Huazhong University of Science and Technology Press. 2012.

https://doi.org/10.1515/9783110573275-010

Index

https://doi.org/10.1515/9783110573275-011

www.ingramcontent.com/pod-product-compliance
Lightning Source LLC
Chambersburg PA
CBHW081101220326
41598CB00038B/7182

9783110573268